THE ART OF
WELDING

THE ART OF WELDING

Practical Information and Useful Exercises for Oxyacetylene and Electric Arc Welding

W. A. VAUSE

FOX CHAPEL
PUBLISHING

ISBN 978-1-4971-0199-9
Library of Congress Catalog Number: 2021945626

To learn more about the other great books from Fox Chapel Publishing, or to find a retailer near you, call toll-free 800-457-9112 or visit us at *www.FoxChapelPublishing.com*.

We are always looking for talented authors. To submit an idea, please send a brief inquiry to acquisitions@foxchapelpublishing.com.

Printed in China
First printing

Because working with metal and other materials inherently includes the risk of injury and damage, this book cannot guarantee that following the instructions in this book is safe for everyone. For this reason, this book is sold without warranties or guarantees of any kind, expressed or implied, and the publisher and the author disclaim any liability for any injuries, losses, or damages caused in any way by the content of this book or the reader's use of the tools needed to complete the processes presented here. The publisher and the author urge all readers to thoroughly review each process and to understand the use of all tools before beginning any project.

Preface

The art of welding wrought iron has been practiced by village smiths in their forge fires for centuries; many beautiful examples of their work remain, most commonly in wrought iron gates which can be seen in almost all historic buildings in many countries. Gas and electric arc welding are, however, comparatively recent developments and in the Introduction their history is briefly outlined.

This book (The Art of Welding, for I maintain that it *is* an art) does not set out to be an exhaustive treatise on the subject but more of a useful discussion of a process ever more widely used and developed in industry, in an easily read and understood form for the novice, who may never actually have seen a welding blowpipe or electric arc being used. It is thus written in as nontechnical a way as is consistent with clarity, steering clear of jargon as far as possible.

Welding has by no means reached the limit of its possibilities, and indeed it could be said that its application to engineering and similar spheres of activity is still in its infancy. I venture to predict that it will be still more widely used in many ways in the future. However, the present purpose is to help those who are strangers to welding but for one reason or another have become interested; this might include those who have just started a course in practical welding, or possibly someone from management who needs to acquaint himself with the broad outline of the subject, or even the amateur who wishes to follow up this absorbing activity with an eye to its creative metalwork possibilities or its application to model engineering.

Some readers may have already have complete gas or arc welding outfits, or may be about to purchase equipment, and to these I would say please, please read the notes on Safety. Welding is fascinating but it does involve certain hazards and risks can be reduced or avoided by such simple steps as always having a fire extinguisher handy, having the key in position on the acetylene cylinder and similar common sense precautions.

In preparing the book I have been greatly encouraged by the assistance of the Engineering Industry Training Board, who kindly consented to the use of a number of the excellent illustrations from their basic manuals, and the Murex Division of B.O.C. Limited who have helped with other illustrations.

Hove, 1984   W. A. Vause

5

Contents

Introduction

In 1809 Sir Humphrey Davy first achieved an electric arc, using two platinum electrodes, but it was not until some 83 years later than two Russian scientists, Nikolas Von Bernados and Stanislav Olczewski, patented the first process of arc-welding, using carbon electrodes in combination with filler wire, a method which with some modifications is still in use for certain purposes today. In 1892 Nikolas proposed the use of bare-wire electrodes, which were in fact used for many years. This method, however, held serious drawbacks, one of which was that only D.C. current could be used for bare wire.

Another serious drawback was that nitrogen from the surrounding atmosphere became absorbed in the weld metal, forming iron nitride, which caused the weld metal to become too hard and brittle. Also atmospheric oxygen created oxidation of the weld deposit. Some means had therefore to be found of obviating these problems, and in 1907 a Swedish engineer, Olgar Kjellberg, patented a very thinly-coated mild steel wire electrode, which vaporized minerals surrounding the arc, thus shielding it from the effects of the atmosphere. However, it was not until about 1912 that the first heavily-coated electrodes were patented by Arthur P. Strohmenger in the U.S.A.

Later other numerous and diverse coatings came to be used, for example asbestos, with an aluminum wire coiled around it. Also sodium silicate and other various chemical substances were used to coat the electrodes, all having the object of creating a gas, or vapor shield, to exclude the air from the arc and the molten metal, and at the same time producing a fusible slag covering the surface of the weld deposit which on cooling could easily be removed.

In 1895 the oxy-acetylene flame was first used by Henri Louis Le Chatelier, a French chemist. Today, however, about 65% of all welding is carried out by the arc-welding process.

PART ONE
Chapter 1

Welding by the Oxy-Acetylene Process (Gas Welding)

The equipment consists of:
The Blowpipe
Two Regulators
The Canvas-Rubber Hoses, plus of course the
Two cylinders of gas, one of oxygen and one of acetylene.

THE BLOWPIPE

There are several makes of gas welding blowpipes on the market, perhaps the best known being the 'Saffire', which is a very efficient, general purpose blowpipe, and is supplied with detachable nozzles of different sizes, numbered as follows:
1, 2, 3, 5, 7, 10, 18, 25, 35 and 45.
These numbered nozzles indicate the approximate consumption of gas in cubic feet per hour at the appropriate pressures.

THE REGULATORS

There are two of these. The Red (or maroon colored) one is the acetylene regulator, and the Black one is the oxygen regulator. Each regulator carries two gauges, one to indicate the contents of the cylinder at any given moment, and the other to indicate the working pressure in lbs. per square inch, which is regulated by the turnscrew provided. This is turned in a clockwise direction to increase the working pressure, and anti-clockwise to reduce it.

Before screwing the regulators on to the cylinders, any dust or other foreign matter which may have collected in the socket of the cylinder should first be blown out by 'snifting' the cylinder momentarily. This is done by opening the valve of the cylinder for a fraction of

(continued page on 10)

Fig. 1 *The Saffire 3 blowpipe.*

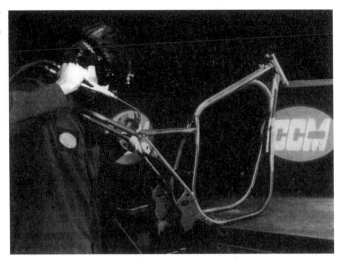

Welding a motor-cycle frame at Clews Competition Machines, where both gas and arc welding are used.

Notes on Safety – Gas Welding

Always store gas cylinders well away from any source of heat.

Make sure there are no leaks from any cylinders, and no grease or oil on or near them.

Keep flame of welding blowpipe well away from cylinders.

Always keep the cylinder key in position so that gas can be turned off immediately.

Always wear goggles, with the correct specification lens, also gloves, or gauntlets.

Wear suitable clothing and leather apron, plus protective helmet if doing overhead welding.

Wear goggles with clear lenses when chipping or grinding.

Make sure there is adequate ventilation, also extractor fans wherever possible, to dispel toxic fumes.

Take particular care when working in a confined space, e.g. tanks, boilers, or drums. Never weld petrol tanks, or similar vessels, until they have been cleaned by steam jet. Wear mask if any dangerous fumes are present, e.g. when welding or brazing galvanized workpieces.

Other workpieces to be welded often have to be de-greased first, by being treated with trichlorethylene, or similar. These operations must be carried out well away from all welding activities.

Never allow caustic soda to come into contact with trichlorethylene, etc.

Always have adequate fire extinguisher handy.

Never use low pressure regulators, or welding blowpipes, on high pressure systems. (Note – this book deals only with the high pressure, dissolved acetylene system).

Stand well away from cylinder regulators when turning on gas, as the glass faces of these have been well known to splinter if cylinders are very suddenly opened.

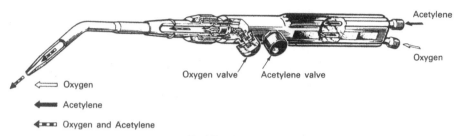

Oxygen valve   Acetylene valve

⟨▭ Oxygen

◀▬ Acetylene

◀■□ Oxygen and Acetylene

Fig. 2 *Blowpipe details.*

a second, and closing quickly with the special key provided.

It will be noted that while the oxygen regulator has the normal right-hand thread, the acetylene regulator has a left-hand thread. The reason for this is so that they can be distinguished even in darkness, and so that there can be no possible danger of confusion. After screwing regulators on, make sure that they are securely tightened up.

Never at any time allow oil or grease to come into contact with any part of the equipment. The reason for this is that oil or grease when mixed with oxygen cre-

ates a chemical combination from which heat can arise which could cause an explosion. Also, while on the subject of safety, never allow the oxygen cylinder (or, indeed, any cylinder) to become exposed to heat, from any source, as this results in expansion, which again could lead to an explosion.

THE CANVAS-RUBBER HOSES

These carry the gases from the regulators to the welding blowpipe, the red of course being for acetylene and the black for oxygen. They are supplied with union nuts at each end, and again

Fig. 3 *Oxygen and acetylene pressure regulators*

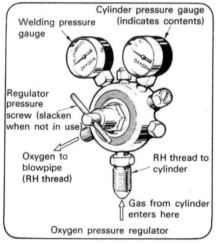

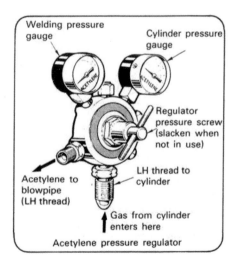

the one for acetylene has a left-hand thread, while the oxygen one has a normal right-hand thread.

With these connected up to the appropriate regulators, and to the welding blowpipe, the assembly is complete. The equipment should now be tested by turning on the gas at each cylinder with the special key, which, by the way, should be left in position on the acetylene cylinder, so that it can be turned off immediately in case of emergency.

LIGHTING THE FLAME

With the working pressure set at about 5lbs., and with a number 2 nozzle in the blowpipe, to light – open the two valves on the shank of the blowpipe (red for acetylene and black for oxygen), opening the acetylene one slightly first with just a small touch of oxygen, then light up, preferably using a flint spark lighter, but is this is not available, having a lighted candle handy on the bench, making sure of course to keep the hand well away from the flame and the nozzle pointing in a safe direction.

Lighting the flame can be, until one gets accustomed to it, a little tricky. Sometimes a backfire may occur. This is due to one or both valves being insufficiently open (usually the acetylene), according to the size of the nozzle. It is better to have the acetylene valve well open, even a little too much rather than too little, as if the latter, all that will happen is that the flame will 'jump' away from the tip of the nozzle, leaving an airspace between it and the tip of the nozzle, whereas a backfire can be a little dangerous. If a backfire does occur, to be on the safe side, it is best to plunge the blowpipe into a bucket of water immediately, especially if the shank feels to be getting hot. (I have known blowpipes to become so hot as to melt the brass of the shank, as a result of a backfire, which causes a flame inside the chamber, but with a modern blowpipe like the 'Saffire', it is almost impossible for this to happen).

If, however, you have opened the valves a little too much, all that will happen is that the flame may jump away from the nozzle, in which case all that is necessary is to close the valves again to shut off the flame and start again.

ADJUSTMENT OF THE FLAME

Having got the flame going fairly normally, i.e. reasonably under control, and with a fairly equal mixture of both gases, on looking at the base where the flame emerges from the nozzle, a small inner cone-shaped flame will be observed. This inner white cone of flame is acetylene, and it may appear as a quite long white feather, or may be quite small. It is very important to get this inner white cone of flame correctly adjusted.

We will assume for the moment, purely as an experiment, that we are about to weld some mild steel, say $1/_{16}$in. (1.6mm) thick, using a No. 2 nozzle, with gas pressure set at about 4lbs. on each regulator. For this, it is vitally important that we have an absolutely neutral flame. The importance of this cannot be exaggerated, for if, in the case of mild steel, an excess of oxygen is present in the flame, this would cause oxidization, in other words, burning of the metal. An excess of acetylene causes carburization, which means that too much carbon is absorbed into the weld, causing the metal to become too hard and brittle.

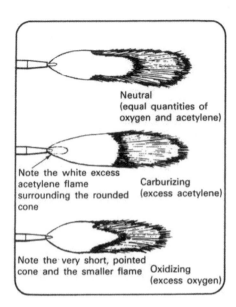

Neutral
(equal quantities of
oxygen and acetylene)

Note the white excess
acetylene flame
surrounding the rounded
cone

Carburizing
(excess acetylene)

Note the very short, pointed
cone and the smaller flame

Oxidizing
(excess oxygen)

Fig. 4 *Flame types.*

To obtain a neutral flame, proceed as follows:

Having lit the flame, continue to turn on more oxygen until the central acetylene flame appears as a white feathery inner cone. Now we come to the finer and final adjustment. Continue to increase the oxygen supply very gradually and carefully, until the white feathery haze on the inner cone just disappears, leaving the white inner cone sharply defined. Great care must be taken not to turn on too much oxygen, only just sufficient to cause the white haze of acetylene to disappear; the slightest amount more will produce an excess of oxygen, which will oxidize the weld metal.

Alternatively, the final adjustment may be made the other way round, i.e. by turning *off* the acetylene until the white haze at the edges of the inner white cone just disappears, taking the same care not to overdo this. Incidentally, it is advisable to check that the flame is still perfectly neutral after welding for a few moments, because as the nozzle gets hot it tends to expand, and therefore alters the working pressure.

Another even more accurate guide as to whether the flame is absolutely neutral or not is obtained by observing the appearance of the weld metal itself while it is in the molten state. The telltale signs are as follows:

In the case of an excess of oxygen, the pool of molten metal presents a burned-up cindery appearance, with the accompaniment of a continuous shower of sparks. An excess of acetylene is even more easily distinguishable. The pool of molten metal then presents a spotted appearance, not unlike strawberry jam, with a lot of pips clearly visible, floating about the surface of the molten metal.

However, with a perfectly neutral flame, the pool of molten metal should present a beautiful, smooth, golden color, which could be likened to melted butter, and should flow easily and smoothly, and with few sparks flying.

It should be borne in mind, however, that the above refers only to the welding of mild steel; other metals require different techniques as we shall see later on.

Always extinguish the flame by turning off the acetylene valve first.

FUSION WELDING

Having arrived at this point, i.e. having become fairly efficient and well practiced in the art of obtaining a perfectly neutral flame, we have arrived at the stage where we can proceed to try a little experimental practical welding, practicing what might be called 'fusion welding'.

This is very good practice indeed for the novitiate in the art of welding and extremely valuable in getting used to handling the blowpipe and developing the skill and technique so highly necessary.

'Fusion welding' is simply practicing the making of straightforward 'beads' of weld along a straight line, on the metal surface, without using any filling wire, and must not be confused with welding proper. For our purpose we will use a small piece of sheet metal (mild steel)

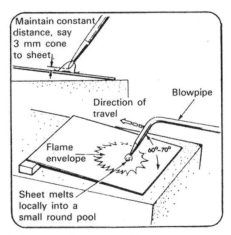

Fig. 6 *Move blowpipe to left once melt occurs.*

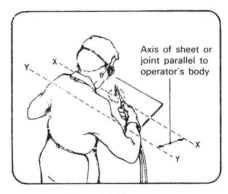

Fig. 5 *Prop workpiece on firebrick etc.*

say 6 × 6in. of 16 gauge thickness set up clear of the bench, preferably on a firebrick. Fitting a No. 2 nozzle and setting the working pressure gauge to 4lbs. on each cylinder, and donning gloves and goggles, now light up the flame, and having carefully adjusted it to neutral, apply the flame to the metal.

Now the exact position of the flame in relation to the metal is very important.

The blowpipe should be directed to the surface of the 'parent' metal (i.e. the metal on which we are about to carry

out the welding operation) at an angle of roughly 45°, but this varies enormously according to the particular job in hand, and as the welding operator becomes proficient his experience will tell him the best angle to suit the particular working conditions applicable at the time.

The blowpipe should be held in such a position that the tip of the *inner* white cone is just clear of the surface of the

Fig. 7 *Adjust speed to pool formation.*

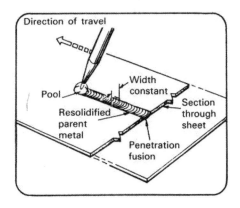

13

metal without actually touching it. The reason for this is that this is the hottest part of the flame, and therefore the most efficient position for quickly obtaining the necessary heat to create a pool of molten metal.

In actual fact the trainee welder should endeavor to develop such skill in handling the blowpipe and hold it in such a way that he will be able to twist it round in his fingers so that the flame can be lifted clear of the pool of molten metal instantly when necessary, momentarily, to avoid burning, or even developing a hole (or 'crater', to use the more technical term). When this happens, very swift action indeed has to be taken, so the student will appreciate the necessity of holding the blowpipe in such a way that the flame may be whipped off immediately the first sign of a crater developing is observed. This skill in handling the blowpipe is only attained by long practice and the learner should not feel too discouraged when a crater forms, as it almost inevitably will in one's early efforts. These things happen to the best of us in the process of learning a new technical skill, or craft, and the only way to achieve proficiency in any art is by meeting these difficulties and overcoming them by constant practice.

However, having thus been forewarned, let me hasten to re-assure the student that this need not be a disaster, and here is the way to deal with it.

If you have lingered a little too long, or have got too big a flame, and the parent metal is overheating, and a crater looks like developing, whip off the flame immediately by a quick flick or twist of the wrist and fingers for a second or two, just to allow the metal to cool a little, and then proceed normally. If, however, you are just a little too late,

and a crater does develop, don't worry, just pass over it and go on welding from the further edge of the hole, and proceed normally to the end of the weld. In the meantime, the crater will have cooled, and the operator can then come back to this later and deal with it at his leisure.

More about 'fusion welding'. As I explained, this is purely for practice, and consists in running beads of weld, without using any filling wire, along the surface of the metal. Keep on practicing this, all the while getting more and more used to handling the blowpipe, getting a perfectly neutral flame, and controlling the pool of molten metal. After a good spell of practice in this, the student should be able to produce nice straight runs, or beads of weld, neatly and evenly, and all about the same depth of penetration. In this connection it will be noticed that the molten metal forms itself in small waves, and this is the appearance each bead should present, when properly executed, nice straight neat even runs, as shown in Fig. 7. To proceed with fusion welds, start at the right-hand edge of the steel plate, and hold the flame steadily concentrated on one spot – just clear of the edge, and hold it there until the metal begins to melt, and in another moment or two we will observe a nice round pool of molten metal beginning to develop. When we have got this pool of metal nicely and fully developed, we can start to move slowly forward, keeping the tip of the inner white cone of the flame just clear of the surface of the plate and keeping the pool of molten metal about the same width and depth all the time. Keep steadily moving forward, the rate of progress being governed by the rate at which the metal melts. After plenty of good practice in this, the student can proceed to something more interesting, i.e. practicing beads of weld using filler wire. For this, we proceed in exactly the same way as we did in our fusion welding practice, but with the addition of filler wire.

Filler wire consists normally (for the welding of mild steel) of soft iron wire, and is usually copper-coated. The purpose of the copper coating is to help absorb any excess oxygen. It is obtainable in several sizes, as in the table.

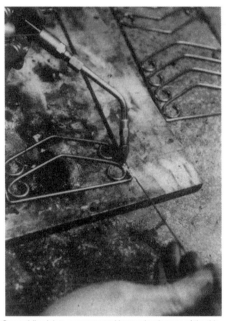

Gas welding joints on ornamental ironwork using one of the thinner filling wires. Note that the operator is left handed.

COPPER-COATED FILLER WIRE

Nom. inch	Approx. inch	Approx. millimeters
$1/_{32}$*	.031	0.8
$3/_{64}$	.048	1.2
$1/_{16}$	.063	1.6
$5/_{64}$	.078	2.0
$3/_{32}$	.094	2.4
$7/_{64}$	.125	3.2
$3/_{16}$	.188	4.8
¼	.250	6.4

Normally supplied in 36in. (910mm) lengths

(*Sometimes less easy to obtain).

PRACTICE WELDS WITH THE USE OF FILLER WIRE

Using the same material as in our fusion welding practice, i.e. 6 × 6in. × 16 gauge mild steel, with a No. 3 nozzle this time, and about 4lbs. working pressure on both acetylene and oxygen, take a $3/_{64}$in. welding wire in the left hand (incidentally, it must be understood that all instructions given refer to right-handed welders only, left-handed welders will have to adapt these notes accordingly) and proceed exactly as before in fusion welding, starting at the right-hand edge of the piece of sheet metal, and proceeding in a left-hand direction. The only difference is that this time having developed a good pool of molten metal, and with the welding wire in the left hand, keep dipping the wire into the pool of molten metal as the weld proceeds, not keeping the wire immersed all the time, but rather using an intermittent dipping motion, the object being to maintain the level of the molten metal in the pool up to the sectional thickness of the material.

Now here may I say a word about 'Penetration'. This is a very important factor in welding, and one of the things

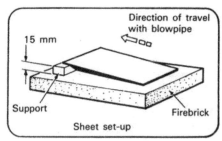

Fig. 8 *Supporting the practice piece.*

reverse side as on the welded side. That, however, will come later. The beginner cannot hope to achieve that ideal just yet. For the moment we must concentrate on practice welds, with the use of wire as described above.

Continue to practice these first simple straightforward beads with the flame held quite steady and the wire preceding

which the welder must bear in mind and be concerned about all the time. As regards purely practice welds, we need not be too worried about penetration, but when we arrive at welding proper, the ideal to be aimed at should be to make sure that the weld has penetrated through the whole material being welded, and in the case of sheet metal, the weld should show through to the other side. In other words, on inspection the weld should be clearly visible on the reverse side, and should, ideally, present almost exactly the same appearance on the

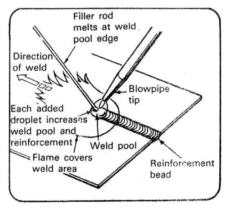

Fig. 10 *Rate of travel needs practice.*

Fig. 9 *Filler rod not in the flame center.*

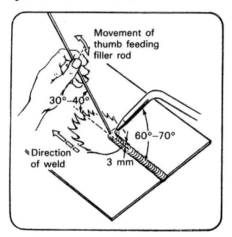

it, dipping into the pool of molten metal at regular intervals, as the weld proceeds. On reaching the end of the run, extra care must be taken to avoid burning the metal, as a crater could quickly develop here on the edge, because of the heat being so greatly concentrated in one small area. So what we do here is to 'weave' the flame from side to side a little, so as to spread the heat over a wider area and disperse it as much as possible, also at the same time running a little more wire into the weld, to build up the edge to normal.

In continuing to practice these welds lies the surest way to success in achieving the objective we are aiming

16

at, namely, to produce sound, 'homogeneous' ductile welds, which are 100% sound and reliable throughout.

The main purpose of these practice welds is to impress upon the mind of the student the vital importance of melting the 'parent metal' which cannot be too strongly emphasized. It is utterly useless and futile merely to melt some welding wire on to the surface of the metal; that is not welding at all. A true weld to be 100% throughout must penetrate right through the entire thickness of the material being welded; in other words 100% fusion must be achieved. Incidentally (and purely in passing, because the novice cannot be expected to reach this stage until after long practice) it can be noted that in order to pass the official A.I.D. Test, i.e. 'Aeronautical Inspection Directorate Test' which the welder must take before being allowed to work on anything connected with the aircraft industry, the first point they will be concerned about is – 100% penetration.

Chapter 2

Exercises with Mild Steel

EXERCISE I. A BUTT WELD ON 16 GAUGE MILD STEEL SHEET

Having become thoroughly practiced in producing the experimental fusion welds described above, the operator can now pass on to something a little more advanced, namely, an actual welded joint of two sections of mild steel (from now on this will be referred to as M/S which is the common practice in machine drawings). A butt weld is a welded joint in which the two sections are placed edge to edge and fully fused together by means of welding to form one complete whole.

In this exercise then, two pieces similar to those we used for our practice fusion welds, i.e. two pieces of M/S sheet about 16 s.w.g. × 6 × 6in., will be needed. These, as before, should be set up preferably on a piece of fire-brick, if available, but the important thing is that they must be raised clear of the bench. This is essential, because if not the bench, being steel (obviously a wooden bench is absolutely 'out') would conduct the heat away, thus greatly chilling the work, and so defeating your efforts to raise the metal to a welding heat. An ordinary building brick could

be used as a last resort, but this can be a bit risky, and is liable to chip and fly under the influence of the heat, which is very great. Alternatively, small steel supports could be used for this purpose, but clearance space must be left underneath the actual line of the seam to be welded. Also needed is some kind of holder or rest on which to support the blowpipe when not actually welding, without having to extinguish the flame. There are several devices on the market specially designed for this purpose, called Economizers, which, as the name implies, also economize in gas. But as a makeshift expedient, just a piece of sheet metal, bent to 90° and screwed to the bench, with a slot cut in the upper part to hold the blowpipe, will suffice; this saves continually putting the flame out and lighting again. Incidentally, the temperature of the oxy-acetylene flame is approximately 3000°C, and the melting point of M/S is approximately 1450°C; the melting points of other metals will be dealt with in later chapters. It is important now to get the two pieces of metal nicely lined up together, with the edges exactly parallel, and level with each other, and leaving a gap of about $\frac{1}{32}$in. (.8mm) between. This done, we are now ready to start the weld.

Again using a No. 2 nozzle, with the working pressure set at 4 – 5lbs. on each of oxygen and acetylene, light the flame, adjust to neutral, and taking up a length of $^3/_{64}$in. (1.2mm) wire, and donning gloves and goggles, all is ready.

The first part of the operation is to 'tack' the two parts together. Tacking plays a very important part in welding. This is done in order to hold the work together temporarily, to enable the welded joint proper to be carried out later. A tack is simply a small, what might be called a 'miniature' weld, and in the case of the weld at present being discussed, the tacks should be kept as small as possible, i.e. the merest touch of wire in the flame, just sufficient to bridge the gap and fuse the two edges together. The first tack should be placed right in the center of the seam, or joint, and the second one about two inches to the right of the first, the third two inches to the left of the first, and so on, following this procedure all the way along the seam to be welded. The reason we use this procedure in tacking is in order to avoid distortion. This is my first reference to the dreaded word 'distortion', but the reader may rest assured that it will not be the last. To explain a little further; if we started at what seems at first acquaintance to be the natural way, i.e. at the right-hand edge of the metal plates, we would find as we tacked along the seam from right to left that the two edges would begin to close inwards toward each other, until before long they would actually be overlapping and over-riding each other. The cause of this distortion is the simple fact that iron or steel when heated ex-

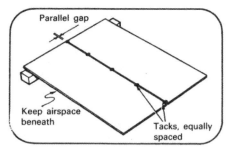

Fig. 11 Place tacks symmetrically.

pands, and later, when cooling takes place, contracts. Now it is this *contraction,* or shrinking, which is one of the biggest problems with which the welder has to cope, and, as may be gathered, is the cause of not a few headaches. Actually, this great natural force can be used by the skilled welder to work *for* him, instead of against him, if he uses his skill and experience to the best advantage. Once again, this knowledge can only be gained by actual experience – all the text-books in the world cannot take the place of this. A book can explain the principles which govern this phenomenon and start the learner off on the 'right lines', so to speak, but from then on, experience alone must be the best teacher.

To return now to our first tack. In making this tack, what we are actually doing is fusing a small portion of our filling wire with the parent metal, which in this case, is our two pieces of sheet metal. As soon as these – wire and parent metal – are fused together, we withdraw the flame from the work, and as soon as we have done this, what happens? The tack immediately begins to cool off, and as it does so, the weld metal in the tack very quickly contracts, and in doing so, it draws together the two edges of the pieces of sheet metal.

Now, if we were to proceed in what seems to be the natural way, namely, to place the next tack about an inch further along, the effect of this second tack would be to draw the edges still more closely together, and the next one still more so, and so on with each successive tack, until, long before we reach the end of our weld, the two edges have started to over-ride, and very soon are actually overlapping. If we were to continue along the seam in this way, by the time we reached the further end our two pieces of sheet metal will have become so hopelessly distorted and out of shape that the whole work would be useless. Therefore, instead of starting the first tack at the edge of the material, we place the first tack in the *center* of the seam. Then by placing the tacks alternatively on each side of the center the heat is spread evenly along the seam, and the expansion and contraction of each tack is balanced by its opposite the whole way along the seam. When

completed, we have our two pieces of sheet metal perfectly tacked together, quite straight, and in perfect alignment, ready for welding.

Just one more point in regard to the above, it is important that the flame should be 'whipped clear' of the metal plate after completion of the tack, in order to keep the plates as cool as possible, thus limiting the distortion as much as possible. In other words, as soon as each tack is placed, lift the flame clear of the plate immediately, thus allowing it to cool before proceeding to the next tack.

Now to complete the weld proper. First, reverse the plate in order to carry out the weld on the opposite side. Beginning at the extreme right-hand edge (a left-hand operator must, of course, reverse this procedure) of the seam apply the flame exactly as when tacking, taking care not to burn the edge, bearing in mind that the exposed edge of the plate will obviously melt much quicker than the material further in from the edge. Establish a pool of molten metal as before, and then feed in the filler wire. Once the pool of molten metal is established, the operator proceeds along the line of the seam, the wire held in the left hand preceding the flame, the tip of it just dipping into the pool of molten metal at regular intervals as the weld proceeds, the flame pointing exactly along the line of the weld.

There are several different 'techniques' or methods of welding; one already hinted at is the 'weaving' method, in which the flame is moved from side to side as the weld progresses, and so on, but this will be dealt with later. For the present, the beginner need only concern himself with a simpler method. That is, to hold the

Fig. 12 *Leftward welding in progress.*

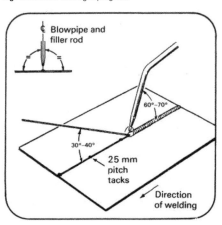

20

flame unwaveringly in line with the seam, while moving forward along it from right to left, keeping pace with the pool of molten metal, and feeding the wire into it as required to maintain the level of the metal up to that of the rest of the material. There is no point in building it up higher than this, as any more will be superfluous and probably only have to be ground off; it will not make the weld any stronger.

Another point is the speed of travel. This is governed by the rate at which the metal melts, and as the weld proceeds, it will be found that the heat will gradually build up and increase, so that the speed of travel may have to be increased slightly. On reaching the further end, it may be necessary to lift the flame off momentarily in order to avoid actual burning, and in fact it may be found necessary to run a little extra wire into the weld at this point, to build up the edge to normal. A further point too, here. After welding for a minute or so, obviously the nozzle, becoming heated, expands, so it is as well to check the flame soon after starting, to make sure it is still absolutely neutral.

In carrying out this weld, the operator's object should be to penetrate the material as deeply as possible without actually burning the metal and causing a crater to form. If this does happen, and it is bound to do so in the beginner's first attempts, do not panic, but as mentioned earlier, simply leave it and continue with the weld; you can always come back later, when it has cooled off, to repair it.

On completion, the weld should present an appearance of a series of evenly dispersed fine waves, the same width the whole way along the seam, and to be passed as a perfectly 'Homogeneous Ductile Weld' the underside should present an exactly similar appearance to the welded surface with 100% penetration right through the material. However, this is something the beginner cannot hope to achieve at this stage; this desirable result can only be attained after considerable practice and experience. Therefore, the thing to do now is to continue practicing these experimental butt welds on sheet metal of varying thicknesses as much as possible. With constant practice of this kind, the operator should eventually be able to achieve the ideal.

EXERCISE II. SINGLE VEE BUTT WELD

After plenty of practice on the above types of welds on sheet metal, the operator should now be ready to pass on to attempt a *single* vee butt weld on thicker material, say ⅛in. M/S.

Edge Preparation. To butt weld two pieces of this material by the oxyacetylene process (commonly called simply 'Gas Welding') it is necessary to bevel the edges (or 'chamfer' them, to use the normal term). This is done so that when the two edges are placed or 'butted' together they form a V-shaped groove. The purpose of this V is to allow greater penetration, in fact, without the V, 100% penetration would be impossible in thicker material. Actually, anything up to ⅛in. (say 3mm) in thickness *can* be welded without any V or any edge preparation, by simply using a small gap of about $1/_{16}$in. (1.6mm) between the edges, which acts instead

Fig. 13 *Single vee edge preparation.*

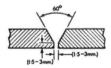

21

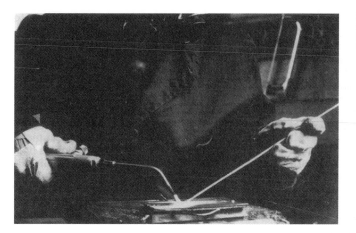

of the V. However, this is not always possible because of limits in dimension, etc. Therefore, for the purpose of our exercise, we will prepare the edges of our ⅛in. M/S plate for a single vee butt weld by filing or grinding them to an angle of 30° on each edge. But we do not chamfer (or bevel) them to a fine knife-edge. We stop short of that, and just leave a slightly blunt 'nose' on each edge. Thus, when the two edges are lined up together, they form a V of 60° angle for most of their depth, but with the bottom fraction vertical-sided.

The weld itself is carried out in the same way as in the following exercise. Generally speaking, in these days, anything over ⅛in. (and, indeed, much thinner material as well) is welded by the electric arc process which is approximately six times faster, and involves less distortion, because the heat though of greater intensity (the temperature of the electric arc is approximately 6000°C) is far more localized and confined solely to the area of the weld, whereas in the case of oxy-acetylene, the heat tends to be diffused over a wide area.

EXERCISE III. A SIMPLE DOUBLE VEE BUTT ON MILD STEEL

For this, proceed as follows.
Setting up the plates in exactly the same way as in Exercise I, proceed to tack in exactly the same way, except that in this case we shall need a slightly larger nozzle, a No. 5; also slightly larger filler wire, say $1/_{16}$in. (1.6mm) or $5/_{64}$in. (2mm) and a slightly increased working pressure, to about 5lbs. on both oxygen and acetylene, again making sure, of course, that the flame is absolutely *neutral.* Having completed tacking, the technique to be followed is exactly the same as in Exercise I, except that it will now be found necessary to use the 'weaving' method, in order to achieve full fusion right across the

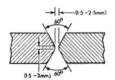

Fig. 14 *Double V butt weld.*

whole width of the vee. In this, the flame is moved across from side to side of the V while the filler wire may be held still while preceding the flame, or it may be moved across the vee from side to side alternately with the flame.

When the weld is completed, the underside should present practically the same appearance as the top surface thus proving that complete penetration has been achieved.

However, it is not to be expected that the beginner will be able to accomplish this straight away; quite a bit of practice is needed to be able to do this, so if, on inspecting the underside the weld deposit does not show through, then a 'sealing run' (or 'pass') may be carried out on the underside, to ensure full fusion, and that it is perfectly sealed up. In this connection it should be borne in mind that in many cases a weld may have to be exposed to the weather on outside work and therefore it would be essential that a perfect seal should be achieved on the underside in order to make it impervious to water.

As mentioned earlier, material of ⅛in. (3mm) and over is usually welded by the electric arc method, but if for one reason or another – perhaps because of no electric power, or no welding machine is available – it is found necessary to gas-weld material over 57in. (6mm) then the 'rightward' method is used; that is, the weld is commenced at the left-hand edge and proceeds from left to right, as opposed to the technique used in welding sheet metal. (A left-handed operator would reverse this of course). The rightward method involves a slightly different technique, but is basically the same, except that the flame precedes the wire. However, it is not proposed to include this (Rightward Welding) or more than a brief

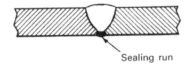

Sealing run

Fig. 15 *Sealing run (after main weld).*

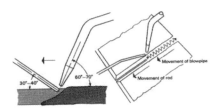

Fig. 16 *Weaving technique.*

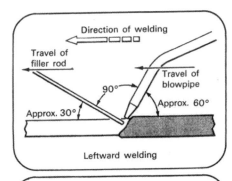

Direction of welding

Travel of filler rod

Travel of blowpipe

90°

Approx. 60°

Approx. 30°

Leftward welding

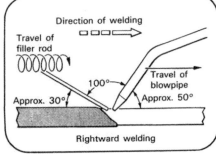

Direction of welding

Travel of filler rod

Travel of blowpipe

100°

Approx. 50°

Approx. 30°

Rightward welding

Fig. 17 *Rightward and leftward techniques.*

23

Thickness of metal	Diameter of welding rod	Edge preparation		Speed mm/min	Thickness of metal
Less than 0·9 mm (20 swg)	1·2–1·6 mm ($\frac{3}{64}-\frac{1}{16}$ in)			127–152	0·8 mm ($\frac{1}{32}$ in)
			Leftward welding	100–127	1·6 mm ($\frac{1}{16}$ in)
0·9–3 mm (20 swg $-\frac{1}{8}$ in)	1·6–3 mm ($\frac{1}{16}-\frac{1}{8}$ in)	0·8–3 mm ($\frac{1}{32}-\frac{1}{8}$ in)		100–127	2·4 mm ($\frac{3}{32}$ in)
				90–100	3 mm ($\frac{1}{8}$ in)
3–5 mm ($\frac{1}{8}-\frac{3}{16}$ in)	3–3·8 mm ($\frac{1}{8}-\frac{5}{32}$ in)	80° V 1·6–3 mm ($\frac{1}{16}-\frac{1}{8}$ in)		75–90	4 mm ($\frac{5}{32}$ in)
				60–75	4·8 mm ($\frac{3}{16}$ in)
5–8·2 mm ($\frac{3}{16}-\frac{5}{16}$ in)	3–3·8 mm ($\frac{1}{8}-\frac{5}{32}$ in)	3–3·8 mm ($\frac{1}{8}-\frac{5}{32}$ in)		50–60	6·4 mm ($\frac{1}{4}$ in)
			Rightward welding	35–40	8 mm ($\frac{5}{16}$ in)
8·2–15 mm ($\frac{5}{16}-\frac{5}{8}$ in)	3·8–6·5 mm ($\frac{5}{32}-\frac{1}{4}$ in)	60° V 3–3·8 mm ($\frac{1}{8}-\frac{5}{32}$ in)		30–35	9·5 mm ($\frac{3}{8}$ in)
				22–25	12·5 mm ($\frac{1}{2}$ in)
15 mm ($\frac{5}{8}$ in) and over	6·5 mm ($\frac{1}{4}$ in)	Top V 60° Bottom V 80° 3–3·8 mm ($\frac{1}{8}-\frac{5}{32}$ in)		19–22	15 mm ($\frac{5}{8}$ in)
				15–16	19 mm ($\frac{3}{4}$ in)
				10–12	25 mm (1 in)

EDGE PREPARATION FOR MILD STEEL SHEET AND PLATE

Fig. 18 *Types of corner joint.*

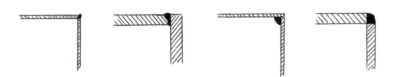

mention of Vertical or Overhead welding within the scope of this book, as it is not intended to be more than an introductory manual to the art of welding for the complete novice.

EXERCISE IV. CORNER WELDS

This is a very common type of weld, and one which the average welder would meet with almost daily in the ordinary course of sheet metal fabrication work as, for instance, to give the most common for example, an ordinary box-like assembly. The form of joint can be a fusion of the edges of two sheets without edge preparation (thin sheet), a 90° butt weld with chamfered edge on the butting sheet (thicker materials), an internal fillet (thin) or an external fillet (thick), Fig. 18. (Fillet welds are the subject of Exercise V). We will take as an exercise just part of an ordinary equal-sided cube-shaped box in 16 gauge (1.6mm) sheet metal with a fusion-welded joint.

In this, the normal method of fabrication would be to fold the main part of the body, i.e. the base and two ends, in the folding machine, and then to join up the two sides to the main body by means of welding. However, for the purpose of a first exercise in practicing corner welds we will first simply take two pieces of M/S sheet 6 x in. similar to those used in our first exercise in butt welds, and use the same size nozzle, No. 2, and working

pressure, 4lbs. of oxygen and 4lbs. acetylene. There are several ways of setting these up in preparation for tacking, one of the simplest being just to stand them up together in the form of an inverted 'V' at an angle of roughly 90°, though the exact angle is not important. The only difficulty with this method is to get the two pieces to stay leaning against each other until one can get a tack on, but with patience it can be done and in the absence of special equipment, it is the only way. However, there are several devices available. One is a special 'corner clamp', and with the aid of one of these the job is made very much easier, as this kind of clamp enables the two parts to be held together vertically edge to edge at 90° until tacking is

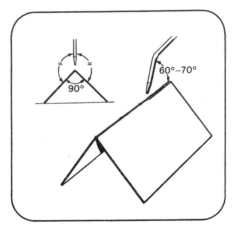

Fig. 19 *Horizontal corner joint.*

25

completed. Even a vise can be a great help if one is available. One part can be held vertically in the vise, and the other held with a pair of pliers or tongs, edge to edge and in alignment at about 90° angle. It is then perfectly easy by applying the flame directly to the corner to fuse the metal at that point, without using any filler wire, to form a tack, and once the first tack is made, the rest is easy. There are also magnetic clamps available which make positioning simpler still.

Once the tacking operation is completed, place the assembled part on the bench in an inverted V position and complete the weld using the same technique as in our practice fusion welds, without using any filler wire, to form a bead of weld along the whole corner joint. It is as well, however, to have ready, in one's left hand, a length of filler wire in case of need, in this case size $^3/_{64}$in. (1.2mm) so that if at any point the weld deposit falls below the full depth required, then the wire is at hand to make up the deficiency. It must be kept in mind that the depth of weld deposit must be maintained the whole length of the joint, consistently and evenly; there must be no weak points anywhere as one weak point could cause a flaw, and the weld would not be 100% homogeneous.

Also, on inspecting the underneath side, the penetration should be visible and just showing through. On the other hand, if the metal is built up too high on the corner weld, this is only causing more unnecessary work in cleaning off, which may result in some variety of uncomplimentary remarks from the person who has to do it – it could be that the welder himself is landed with the job! The above exercise is, of course just an elementary example of a corner weld, but the operator should get plenty of good practice in these until really proficient in carrying them out, because he will find this of great benefit to him in the course of his welding career.

Complete Box Fabrication. Coming back now to our box fabrication, as stated previously, the main body is usually folded in a folding machine, but in many cases a folding machine is not available. In that case, each section will have to be welded separately. To do this, tack the two ends to the base, following the procedure outlined in Exercise III, but do not weld yet. Now lay this tacked section (i.e. the base with the two ends) on the bench, on its edges. Cut a strip of thin sheet metal, and lay it across the top. This is purely to support the next (side) section, which we now lay on the top with the edges of this in alignment with the ends and the base. Make sure that the bottom edge is in perfect alignment with the edge of the base, and very, very slightly overlapping. Now proceed to tack in the usual way, placing a tack about every inch. It helps a great deal to get a tight joint if you keep a light hammer handy with which to tap each tack smartly while it is hot; this draws the two edges really tightly together, thus making the job of welding very much easier. Remember, the procedure is to tack only the base edges first, leaving the ends still open. Then having got the base edges tacked, proceed to tack the ends, starting from the bottom corner, and working toward the open ends. Repeat for the opposite side. It is very important to keep to this procedure, because it is the only way to defeat our old enemy, distortion. This is a good general rule to follow in all welding – to tack up everything first, before commencing welding proper, and in the order I have described. By following this procedure, the forces of expansion and contraction are dissipated and any danger of locked-up stresses is avoided; result, no distortion.

It will be gathered from the above that the process of tacking in the correct order is almost as important as the actual welding itself.

While on the subject, I would like to commend another extremely good method of tacking sheet metal sections which, however, can only be done with the aid of an assistant. In this method the assistant (wearing a pair of light-tinted goggles) holds the two sections together edge to edge and in alignment, using a pair of tongs or pliers in each hand. The starting point is the corner at the welder's right hand and the two pieces of metal are manipulated to bring them together at the exact point at which the welder is tacking at any given moment. Thus, the assistant can move the parts up or down as required for each tack, as the welder progresses toward the open ends of the seam, the assistant watching closely and positioning the two sections together for each tack in turn, while the metal is still hot. This method eliminates all danger of distortion, and greatly facilitates the operation of tacking, but of course it depends solely on the availability of a skilled assistant.

EXERCISE V. FILLET WELDS

A fillet weld consists of a bead (or beads) of weld deposit laid inside a corner made by two components assembled usually (but not necessarily) at 90°. No edge preparation (i.e. chamfering) is required. An outside corner joint (Fig. 18) is a fillet, using the two edges of the metal sheets as the 90°

Fig. 20 *Too much heat, or wrong blowpipe angle, can cause undercut.*

faces of the corner into which the weld deposit is laid.

For this exercise a simple fillet weld between two sheets of metal will be made. Take two pieces of M/S similar to those we have used in our earlier exercises. Set them up for tacking exactly as in Exercise IV for the corner weld (see Fig. 19) and tack up in the same way. Af-

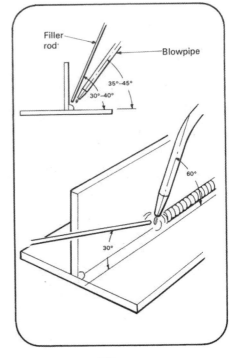

Fig. 21 *Welding a square T fillet.*

ter tacking is completed, turn the work over so that it is set up on the bench in the form of a V right way up, using any odd pieces of metal to support it. For a fillet weld one usually finds that a slightly larger nozzle is required together with heavier filler wire, about $1/_{16}$in. (1.6mm) size. So, fitting a No. 3 nozzle, with the working pressure set at 4lbs. of oxygen and the same of acetylene, and using $1/_{16}$in. copper-coated filler wire, we are ready.

The great danger to avoid in fillet welds is 'undercut'; this is something to be avoided in all welding, but particularly so in fillets. Undercutting is caused by having too big a flame, and therefore too much heat, with too large a nozzle. As can be seen from the illustration (Fig. 20) this causes the area at the edges of the weld to become burned, which obviously creates a weak point in the structure. At the same time there must also be completely 100% fusion all the way along the course of the weld.

Portable gas welding equipment in use to repair a narrow boat's bow rail.

Chapter 3

Gas Welding Other Metals

STAINLESS STEEL

Stainless steel sheet lends itself very well to welding by oxy-acetylene, or, alternatively, it can be brazed. There are various kinds of stainless steel, and a slightly different technique is involved in the welding of this metal, including the use of a special flux, with the correct filler wire, for each particular type of stainless steel. The manufacturers supply all types necessary for niobium, molybdenum or heat resistant types of stainless steels, etc.

As stated in the Preface, this book does not set out to deal in great depth with specialized subjects, such as this for instance, and having regard to the multiplicity of alloys of this and other metals, it will be understood that exhaustive treatment of them cannot be undertaken. The reader is recommended to manufacturers' standard text-books for detailed information on specialized applications, etc., of the various alloys.

For the benefit of present readers, however, I will briefly describe carrying out a simple butt weld on 16g stainless steel sheet metal purely as a practical exercise.

For the purpose, use two pieces of stainless steel 6 x 6in. x 16g and set them up as that for M/S sheet metal, described in Exercise I. The flame must be completely neutral; on no account should a carburizing flame (excess acetylene) be used. With a working pressure of 4lbs. and a No. 2 size nozzle, take a length of $1/16$in. filler wire appropriate to the particular type of stainless steel being welded, heat the end slightly and then dip it into the flux, which will adhere to the end of the wire in the form of a slight tuft. Now tack up in the usual way, then brush the appropriate flux (paste flux is recommended) on the underside of the seam.

The actual technique of manipulating flame and wire is slightly different from that of M/S. In the case of stainless steel, the filler wire is held still while immersed in the pool of molten metal, while the flame is moved in wide sweeps across the width of the seam as the weld progresses from right to left along the seam. Once started, do not on any account stop at any point during the course of the weld, as this may cause cracking; the weld should be completed as quickly as possible for the same reason. Also, on completion of the weld, the flame should be withdrawn slowly to avoid sudden cooling, which could also be a cause of cracking. This is the main thing to watch out for in welding stainless steel, but by withdrawing the flame gradually the metal has time to normalize itself. To summarize, weld as quickly as possible; do not stop during the progress of the weld, and withdraw the flame slowly, passing it over the area of the weld once or twice, allowing the metal to cool slowly and to normalize itself. After completion of the weld, the flux must be removed, partly by chipping, and partly by the use of hot water and a wire brush. Afterward the weld may be ground, buffed or polished, until it is completely invisible.

CAST IRON

There are two main kinds of cast iron, white cast and gray cast. White cast is very hard and brittle, and is therefore not so easy to weld, and also cracks very easily under the influence of heat. Gray cast is one of the easiest of all metals to weld, and being softer, is not so brittle, and therefore resists cracking much better. A third form of cast iron is discussed later in the book.

The technique of welding cast iron is a little different from that of mild steel. Firstly, we use what is called a 'carburizing' flame, i.e. a flame with a slight excess of acetylene. See Fig. 22, in which the lowest one is what is sought. To obtain a carburizing flame, having set the flame at neutral as before explained, we now open the acetylene control valve a tiny fraction more, until a slight 'feather' is visible at the end of the inner white cone. Not too much – just enough to be visible.

The filler rod for cast iron must be high in silicon content, and for this purpose the author recommends the British Oxygen Co's Alda Super Silicon, or for better quality cast iron, Ferrotectic. Also, a special flux is necessary – Cast Iron Welding Flux, also supplied by manufacturers.

Now to get down to the actual welding technique. For experimental purposes we will take a simple straightforward butt weld on good quality gray cast iron, say $1/4 - 5/16$ in. (6 – 8mm) thick-

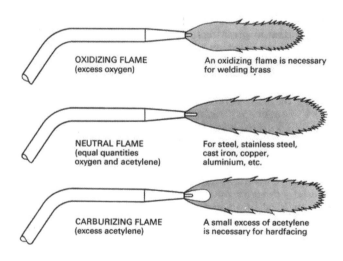

OXIDIZING FLAME
(excess oxygen)

An oxidizing flame is necessary for welding brass

NEUTRAL FLAME
(equal quantities oxygen and acetylene)

For steel, stainless steel, cast iron, copper, aluminium, etc.

CARBURIZING FLAME
(excess acetylene)

A small excess of acetylene is necessary for hardfacing

Fig. 22
Flame types.

ness, which is completely open at the ends, i.e. not part of some other casting, and therefore should have no locked-up stresses and is also perfectly free to expand when heated. The melting point of cast iron is much lower than that of mild steel (the actual figures are: mild steel melts at about 1450°C, and cast iron approximately 1250°C), so a No. 3 will be found adequate for this thickness, setting the pressure gauge at 4lbs. or 5lbs. on both cylinders, and using the suggested filler rod.

To prepare the Material. The edges of the two pieces to be welded should be ground to chamfer angle of 30° on both sides, thus forming a double-V when placed together in alignment. Note that the edges are not ground to an absolute knife-edge, but a small section is left untouched, forming a 'nose'. This helps in the case of a broken casting to line up the two parts, so that they come together exactly as they were before being broken. It is very important to do this in a fractured casting, because it seldom, if ever, breaks in a straight line; if a little of the broken edge is left untouched in the center of the V, then it will be found easy to line up the two parts so that the 'grains' dovetail into each other exactly as in the original casting. Having got them lined up, and using the carburizing flame as described above, pre-heat the whole of the material by playing the flame over the whole area for a few minutes. In a case such as we are at present dealing with, it is not absolutely necessary to bring the material to red heat, as the two parts are perfectly free to expand, and are not locked-up in any way. In the case of a large casting, however, it would probably be necessary to preheat in a spe-

cially prepared muffle furnace, bringing the whole casting to a high and even temperature over all and welding while still hot, then replacing in the furnace to be cooled down slowly.

First heat the end of the filler rod and dip into flux. After sufficiently preheating, the work is now ready for tacking. To do this, raise the leftward end of the joint to a welding heat, and here the operator must be rather careful, because cast iron does not readily show signs of melting until it is in fact almost at melting point. It is therefore necessary to keep a sharp eye open to detect the first signs, and to be ready with the filler rod to insert into the pool of molten metal as soon as it forms. The idea is to keep poking and prodding at the metal with the filler rod until the parent metal is found to be melting, and then the rod must continue to be used with a puddling, stirring motion, while keeping the flame still, and bridging the gap at this point until a fairly strong tack is formed. This puddling action is necessary in order to avoid blow-holes, which are very liable to form in cast iron, because of the various gases which are generated during the welding process. The puddling, stirring action closes up the blow-holes as they form. If this were not done, the resultant weld would be found to be porous. Having now formed a fairly stout tack at the leftward end of the joint, the operator now moves to the rightward end to commence the weld proper. Note that it is not necessary to place a large number of tacks as in mild steel, because in cast iron expansion and contraction is minimal. Proceed with the weld in the manner described, from right to left, stirring and puddling with the filler rod as the weld progresses steadily along to the leftward end of the joint; continue until the weld is completed and full penetration achieved. The flame should be withdrawn slowly, and then played over the completed weld area, to allow it to cool slowly to avoid cracking, for several minutes, gradually withdrawing the flame.

May I repeat, when the weld is completed, great care must be taken in cooling. The flame must not be taken off suddenly, and the idea of playing the flame over the weld is in order to slow up the cooling process as much as possible. All drafts of cold air from open doors or windows should be rigorously excluded during welding of cast iron. If cast iron is cooled too quickly after welding, there is not only the danger of cracking, but the metal will also tend to become hard and brittle, and therefore less machinable. This slow cooling process is known as 'annealing', and if this is properly carried out, the resulting welded joint should be easily machinable, and after grinding and on inspection, no blowholes should be visible.

Afterward, the work should be buried in a bucket of sand or similar material until completely cold.

WELDING ALUMINUM

Aluminum is a beautiful metal to weld, and in some ways one of the easiest, and it lends itself very well to welding indeed. But it has two characteristics which cause some difficulty. One is the low melting point of this metal. Aluminum melts at about 660°C which is very low compared to that of steel or cast iron, for instance. The second characteristic which causes difficulty is the film of oxide which is always present on the surface of aluminum. No amount of filing or wire-brushing can remove this, because it forms on the surface again instantly. It is for this reason that a flux is necessary, to dissolve this oxide film.

The number of alloys of aluminum are legion. (An alloy of any metal is a combination of that metal with other elements). Most of these alloys of aluminum are outside the present scope and will not be dealt with. The author proposes to confine himself solely to pure aluminum, dealing with the welding of this material, which to the uninitiated can prove very difficult. It is my purpose to try and help the beginner over these first hurdles, because it is the first attempts which are the most difficult, and it very often happens that because he feels frustrated, the novice tends to become discouraged. It is with this in mind that the author, remembering his own far off early efforts, proposes to deal with this in rather greater detail, with the object of helping the learner over this critical initial period, because, once over these first hurdles, he will then find it fairly plain sailing, so much in fact, that I can assure him that he will then actually very much enjoy welding this beautiful metal (however wildly impossible that may seem at the moment). First then, let us try a little experimental 'fusion' welding to start with, to get ourselves accustomed to the feel of this metal.

For this purpose, a piece of 16g aluminum sheet metal will suffice, (say about 6 x 6in. in size). Setting the working pressure at approximately 4lbs. on each cylinder, we shall need nozzle size No. 1, some flux (that recommended is 'Alotectic' aluminum welding flux), a few lengths of aluminum wire (the author's practice has usually been to cut a few strips from the original sheet, but manufacturers also supply suitable aluminum welding wire for the purpose), of $1/16$in. (1.6mm) thickness. The correct flame required is a very soft one, with a very slight haze of acetylene showing just a tiny white feather visible at the end of the inner white cone. The object of this is to eliminate all possibility of an oxidizing flame.

Using the same working pressure and the No. 1 nozzle, heat the end of the filler wire (aluminum) and dip the end into the flux. Then, holding the rod upright in a vertical position with the fluxed end upwards, by playing the flame up and down the fluxed end of the rod, it will be found that the flux will liquefy and run down the wire in an even film, thus obviating the necessity of otherwise having to continually be dipping into the flux. A small point here – always keep the Alotectic flux can tightly closed when not in use, as the flux very quickly deteriorates on exposure to the air.

Practice a little of the fusion welding first, exactly in the same way as in the earlier exercises in mild steel, except that in the case of aluminum it is advisable to pre-heat the metal first, by passing the flame over the whole of the surface for a few minutes beforehand. This is because aluminum is such a good conductor of heat that the heat has to spread over the whole surface before it will melt at one particular point. A good method of gauging the temperature of the metal is to simply rub a match-stick, or any small piece of wood, or a piece of soap, on the surface of the metal. If this leaves a brown mark, then we know that the 'ali' is hot enough to start welding. The big difference between ali and mild steel is that, unlike steel or iron, it does not change color with heating, or show any signs of the approaching melting point, except perhaps a slight

darkening shade of gray. This is one of the things which make ali welding difficult at first.

The only way by which the operator can detect the vital moment is to keep a close and unremitting watch on the metal all the time; the first visible sign of the approaching melting point is when the flux starts to run, i.e. changes from a powder to a liquid form, then, a little after, the ali begins to take on that darker shade of gray. Incidentally, have the filler rod in your left hand (the right hand of course in the case of the lefthanded welder), and keep flicking the surface of the metal while waiting for the melting point to be reached, because this is one of the ways which helps the operator to tell when this vital moment is going to occur. By constantly flicking the surface of the metal with the filler rod, the operator is given a clue as to the moment of melting point, and so is ready to plunge the filler rod into the pool of molten metal as soon as this develops. However, at the moment we are only practicing fusion welds, so not much filler metal will be necessary with these. Now, the operator who has never welded aluminum before will almost inevitably find himself causing the metal to collapse into holes or craters at first, and this can be very discouraging, but here let me entreat the welder not to lose heart. All welders have had to pass through this difficult period, but if you persevere success will crown your efforts.

Exercise in Butt Welds. Using say, 16 gauge thickness to start with, cut a couple of pieces of about 6 x 3in. and line them up in the same way as we did with butt welds in mild steel. Proceed to tack in the same way, i.e. the first tack in the center, and then placing tacks about every inch or so, alternately, at the left and the right. Now, starting just clear of the extreme edge of the metal, with the rod already fluxed and the metal well pre-heated, start with the flicking motion already described, and as soon as the flux liquefies, be ready for the moment of melting. Instantly this occurs, dip the rod into the pool of molten metal and proceed to follow along the line of the seam. One of the great secrets of success in ali welding is to start a little away from the extreme edge of the metal, as otherwise if one starts directly on the edge of the metal it immediately burns away, with woeful results. The fully skilled operator will of course avoid burning, and this advice is designed to help the not-so-skilled, and those inexperienced in aluminum welding. The little bit left unwelded at the extreme edge can easily be returned to later and welded when the surrounding metal has cooled. Another secret is in the flicking motion mentioned above. By doing this, not only is the operator more ready to start the weld at the moment the ali melts, but this flicking of the surface with the filler rod helps to break the skin or film of oxide, which, as mentioned earlier, is one of the main bugbears of welding aluminum. Breaking this skin allows the operator to proceed immediately with the weld as soon as melting point is reached. Intense concentration is necessary in these first vital moments, but once started, he will find the metal flowing smoothly and evenly, and quite speedily. As the weld proceeds, and the heat begins to build up, the operator may find that he has to increase the speed of advance progressively as he nears the end of the run, but this depends on how much pre-heating he has carried out beforehand. It will probably be necessary to build up the edge with extra filler wire to avoid a crater.

The weld procedure itself is similar to that required for welding mild steel, except that the operator will find that he will not have to do much weaving, i.e. moving the flame from side to side, but rather keep the flame steady while moving forward, almost drawn along by the flicking motion of the preceding filler wire. So to recapitulate: starting just clear of the edge of the metal and as soon as the aluminum melts holding the flame steadily moving forward, while at the same time dipping the filler wire into the pool of molten metal as required with a sort of 'dragging' action, it will be found that the molten aluminum will form itself into evenly distributed waves similar to that of mild steel.

It is worth stressing that it is inevitable that the beginner will find craters developing with great suddenness, which is very disconcerting, but don't be discouraged; simply pass over that area, leave it to cool off, and come back to it later. Proceed from the far side of the crater and on completion of the run, come back to the damaged part and deal with it at leisure. There are far more advanced techniques and equipment for the welding of aluminum and all its many alloys, such as the argon-arc and CO_2 processes, and of course most large engineering firms and works now use these processes, and have done so since about 1950. Even many small workshops do so, because the great advantage is that no flux is required and this provides a tremendous saving, both in time and labor, as well as the provision of acid baths, etc.

Cleaning off the flux is highly important: it must be thoroughly removed after welding aluminum, as if not it will eventually corrode very severely. To clean off, thoroughly brush the completed weld with a wire brush in hot water, until all traces are removed. This is the simplest method, and indeed, probably the only method available to the average reader, but it is quite sufficient for ordinary purposes, the exception being anything intended to contain or cook foods. These must be cleaned with caustic soda and then nitric acid, followed by thorough washing in clean water.

WELDING CAST ALUMINUM

The technique for this is exactly similar to that of welding wrought aluminum, except that a special filler rod is required, and preferably the special flux for this. It will be found quite easy to weld, using a slight puddling motion with the rod, similar to that used in the case of cast iron. Also as with cast iron, aluminum castings must be pre-heated but must be carefully supported on bars or blocks, so that the weight is evenly distributed, and in no place can the casting sag or droop when heated.

The right temperature for the preheating can be gauged by the melting of a spot of soft solder. When this melts the pre-heating temperature is about right. The flame adjustment for aluminum is the same as for wrought aluminum, with just the merest haze of excess acetylene. This is really to ensure that there is no excess oxygen present. The flame is kept steadily on the weld, with no weaving.

After welding, all traces of flux must be thoroughly cleaned off, to prevent corrosion. Also the casting must be allowed to cool down slowly and uniformly.

WELDING OF BRASS

Brass is an alloy of copper and zinc in various proportions, according to specific requirements. Its melting point is around 900°C (approx. 1652°F). Brass is a very easily weldable metal, the only danger to guard against being the occurrence of blow-holes, which are caused by gases generated by the heat of the welding. To avoid these, the operator uses a puddling action with the filler rod in a similar manner to that used in the welding of cast iron. The best filler rod in my experience is Sif-bronze.

Type of flame This should be slightly oxidizing, that, with a slight excess of oxygen. The purpose of this is that a slight excess of oxygen causes a film of oxide (or skin) to form on the surface of the metal while being welded, which helps to contain the gases, and so to prevent blow-holes occurring. Also, a slightly smaller nozzle size than would be used for an equivalent thickness of mild steel is needed, because brass melts at a much lower temperature than M/S.

Before commencing welding, heat the end of the filler wire in the flame, and then dip this into the can of flux. Continue to do this at intervals as and when necessary. The technique of welding brass is quite simple. Having heated up the metal until a good pool of molten metal is formed, the operator then plunges in the flux-coated filler rod and with a slightly puddling, stirring motion, proceeds in a similar manner to that for cast iron, all the while keeping a sharp look-out for blow-holes forming. It goes without saying that complete penetration must be achieved throughout. When completed, the rough surface can easily be machined away, or even smoothed off with an ordinary electric sander, when all signs of the weld will disappear, and the resulting job will appear as one solid brass formation, with the weld virtually invisible.

WELDING COPPER

Copper lends itself very well to welding, although very strong joints can also be achieved by brazing this metal. As is well known, one of the chief characteristics of copper is its very high thermal conductivity – it conducts heat more rapidly than any other common metal with the exception of silver. Its thermal conductivity is ten times that of lead, six times that of iron or steel, and almost twice that of aluminum. For this reason it is necessary to fit a much larger size nozzle to the blowpipe than would be used for the same thickness in, say, mild steel, and flame adjustment must be absolutely neutral, otherwise porosity (blowholes) may be formed by trapped gases. Before welding copper it must be thoroughly cleaned and degreased, by using a good grease solvent (if possible trichlorethylene).

Edge Preparation For sheet copper of up to 16 s.w.g. (1.8mm) no beveling is necessary, the two edges being treated as a simple butt weld, with a gap of about half the thickness of the sheet or plate. For copper plates of $^3/_{32}$in. (2.4mm) up to 48in. (9mm) a single bevel is necessary, and for plate over this, a double vee is advisable, with a gap of about $^3/_{16}$in. (5mm). Success in welding copper depends a great deal upon the speed with which it is carried out. The edges of the joint should be preheated by the blowpipe for a short time prior to welding and the filler rod of copper should be appropriate to the thickness of the material.

Welding Technique For normal thicknesses of sheet, or plate, the downhand technique is advisable, using the leftward method. Brush flux on both sides of the material, and also on to the rod. The flame is given a slightly weaving motion while the filler rod is kept in the pool of molten metal. It is best to commence the weld at a point about a third of the length of the seam from the end of the joint (i.e. the finishing end), and then weld to the end of the seam. Then, starting from a point about *two* thirds from the end of the seam, continue to the beginning of the first weld, then finish off the remaining part, by starting from the beginning of the joint. This sounds a bit complicated, I know, but it is known as the 'backstepping' method. The idea is to spread the heat so that distortion is avoided or reduced to the minimum. After welding, lightly hammer the weld while still hot. This hardens and strengthens the weld.

Copper welds can be bent up to 180 degrees without any signs of cracking or fracture, and can even be twisted without any signs of damage. They should also be leak-proof, even under high pressure.

Chapter 4

Brazing

Brazing is a very handy and useful method of joining two metals together, its advantage being that it does not require so much heat, and therefore it is useful in cases when too much would be harmful to delicate components, such as fine instrument work. Indeed it is often used in much heavier work, and in quite heavy engineering structures, where there is no particularly heavy stress involved. It can also be very useful in the case of cast iron, as it lends itself very well to this metal, provided it is absolutely clean and bright at the surfaces to be joined. The same applies to brazing generally; the surfaces to be joined must be free from oil or grease and dirt.

The Technique of Brazing. Presuming we wish to unite two metal parts by this method, having got them well cleaned, we proceed to pre-heat the whole of both parts overall. The filler wire recommended is 'Brazetectic' in conjunction with Brazetectic flux. Preheat the end of the wire and dip the end of this into the can of flux, when it will pick up a small tuft of flux, just as in brass welding. Now, having raised the work to a good red heat (just beginning to dull) and using a neutral flame, proceed with a rubbing action with the filler rod, keeping the flame steadily moving forward, but not weaving it. The filler metal will be found to flow evenly along the whole length of the seam, and can be helped along by occasionally sweeping the flame along the line of the seam to encourage the flow of the molten brazing metal, which will form itself into a smooth and even fillet. On completion, the flux can easily be cleaned off by plunging into water while still hot and by chipping. Incidentally, it may be relevant to remind readers that plunging hot brass (and also copper) into water actually softens (anneals) the metal, in complete contradiction to steel, which, of course, hardens when plunged into cold water while still hot.

Brazing is particularly useful in the case of cast iron and malleable castings because of the much lower degree of heat required, which is a big factor in the case of cast iron, since it very much lessens the danger of cracking. But it cannot be stressed too strongly that for successful brazing the surfaces (of cast iron especially) must be absolutely clean and bright. Brazing is also a very useful method of uniting stainless steel, where no great stresses are involved. Here again, the surfaces must be clean and free from oil or grease, and the technique is similar to that described.

Aluminum Brazing Aluminum and its alloys can be joined by brazing. This is not welding in the true sense of the word since the parts to be joined are not fully fused together. Nevertheless brazing is a quick and simple method of joining aluminum or its alloys where the finished components are not subjected to any great strain or stress. Aluminum brazing is only suitable for fillets, T joints or lap joints, not for butt or corner joints. It is carried out with oxy-acetylene and the flame should show a very slight excess of acetylene to ensure that there is no excess oxygen, which is fatal to aluminum. Heat the tip of the filler wire (a strip cut from the original sheet or a length of the correct filler wire) and while still hot, dip it into the flux, picking up a small tuft of flux, then run the flame up and down to form a thin coating of flux as previously described. After tacking in the usual way, slightly pre-heat the whole assembly and drop a spot of flux on to the hot surface near the joint. As soon as this starts to run the temperature is just right. Flick the filler rod along the line of the joint, quickly followed up with the flame, running rod and flame together quite swiftly along the length of the joint. The result will be a beautiful, smooth, even fillet. This can then be cooled in cold water, and thoroughly brushed with the wire brush, to remove all traces of the flux.

Chapter 5

Oxy-flame Cutting

No welding notes would be complete without a section on this. Most big engineering firms employ automatic cutting machines (called profiling machines) of various types. The latest is the electronic photo-electric cell type, in which a photo-electric cell automatically follows the design or drawing required to be reproduced and the oxyacetylene flame follows this, cutting with perfect precision on the mild steel plate underneath. (Cast iron cannot be cut by this method, except with the use of special powders). Obviously the amateur, or the small workshop, cannot enjoy the use of these expensive machines, nor would they have the scope for them. So I propose here to deal solely with hand-cutting methods and equipment.

The principle of oxy-cutting is based on the fact that when oxygen comes into direct contact with hot steel, the steel is immediately converted to iron oxide, and the hand-cutting blowpipe makes use of this. It is generally similar to a welding blowpipe but the nozzle is different, since in addition to the outlets for the oxy-acetylene flame on a welding nozzle, there is a central orifice on a cutting nozzle which directs a jet of pure oxygen on to the heated metal; a release lever for controlling this cutting jet is fitted to the blowpipe.

A typical example of a cutting blowpipe is the Saffire Universal Cutter, which can be used on mild steel of 55in. – 6in. thickness (3 – 150mm) using the appropriate nozzle size. As an example, if it is intended to cut a piece

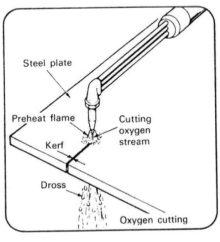

Fig. 23 *Cutting steel plate.*

Steel plate

Preheat flame

Cutting oxygen stream

Kerf

Dross

Oxygen cutting

of 48in. (9mm) M/S, a $\frac{3}{64}$in. (1.2mm) nozzle is used.

Having marked off the line of the cut, it is best to 'bob-punch' this, with an ordinary bob-punch, or at least to scribe the line, because obviously an ordinary chalk line would be burned away under the flame. Donning gloves and goggles, set the working pressure on the pressure gauge to about 25 to 30lbs. (this will in fact cut up to 42in.). Turn on both valves on the blowpipe (i.e. both oxygen and acetylene), but with the acetylene valve fairly widely open and the oxygen valve only slightly open, light in the usual way (keeping the hand well clear) and adjust the flame, just as with an ordinary welding blowpipe, i.e. to a neutral flame; this is the heating flame. The 'Cutogen' nozzle has four small orifices, emitting four small flames, with the cutting jet in the center, but this jet does not come into operation until the operator releases the central lever, which he does only when the metal is hot enough to start cutting.

Apply the flame to the extreme edge of the metal to be cut, holding the inner white cone of the flame almost but not quite touching the surface of the metal. With experience, it is quite permissible to have the inner white cone actually in contact with the extreme edge of the metal, in order to get a quick start. In fact, one usually finds that there is a slight burr or ragged edge on the extreme edge of the metal, where it has been sheared, and this can be used to great advantage, because the burr quickly becomes incandescent, and as soon as this occurs, the metal will start to cut immediately the cutting jet is released. Then as soon as cutting begins, the flame is moved steadily along the line of the cut, quite slowly, at a speed which ac-cords with the progress of the cut. The operator must keep the nozzle absolutely in a vertical position all the time, and move along at a steady regular speed. If too fast, the metal will cease to cut, in which case, he must shut off the cutting jet, heat up the metal again, and start afresh. If a back-fire occurs (which often does happen) this is usually due to small particles of scale, which always forms on mild steel when sheared. This need cause no concern, as all hoses today are fitted with non-return valves which prevent any danger of a blow-back of a flame to the cylinders, which of course could be highly dangerous.

If particularly accurate cutting is required, the operator can use a guide in the form of a piece of angle-iron clamped to the work, or some similar device, but some caution is needed measuring the distance from the guide to the edge of the cutting jet orifice. Care must be taken to measure from the edge of this, nearest to the side of the work which one wants. In other words, one must allow for the diameter of the orifice in measuring the distance from the guide, otherwise one may finish that much too narrow or that much too short.

The secret of good clean cutting is in always keeping the nozzles clean, the cutting jet orifice in particular. This is liable to become clogged, and to keep it clean, the operator should use the special reamers supplied for the purpose. Never use steel wire as this would soon enlarge the copper nozzle orifice of the cutting jet. Effective and accurate cutting can only be obtained if the jet orifice is absolutely clean and sharp and square with the end of the nozzle. The accompanying table gives a brief guide to oxy-cutting of mild steel of varying thicknesses, using the Saffire High

Pressure Blowpipe with the Saffire A-NM Acetylene nozzles.

Other gases can be used in conjunction with oxygen in place of acetylene for cutting purposes, especially in profiling machines, e.g. coal-gas, North Sea (or natural) gas, or propane, butane, and other similar gases. The two last are supplied in liquid form in appropriate cylinders and give off a highly inflammable and volatile gas; either in combination with oxygen is highly effective for oxy-flame cutting purposes, though of no use for welding. They are mostly used in profiling machines, because these hold the flame-cutting jet with absolute rock-like steadiness, maintaining exactly the required uniform speed and distance from the surface of the metal – factors which no human hand can reproduce.

Hints to help in obtaining a perfectly clean cut by Oxy-flame. A perfectly clean cut can only be obtained when the central jet of cutting oxygen emerges from the nozzle absolutely straight and true – like the lead of a pencil – so before commencing actual cutting, a check should be made by simply opening and closing the control lever of the cutting jet momentarily when it will at once be seen how the central cutting jet is behaving. If it emerges from the nozzle in a fanned-out or other distorted form it will not cut cleanly.

The cause of this may be some obstruction inside the jet nozzle which can be cleared by the use of the correct size of the special reamers which are supplied by B.O.C. Limited. On no account must bits of steel wire or suchlike be used, though in extreme cases where the special reamers are unobtainable a suitable size copper wire may be used as a temporary substitute.

The fault may be due to the fact that the edges of the orifice of the cutting nozzle may have become pitted or burned or otherwise damaged. In this case the cure (apart from fitting a new nozzle) is to re-face the tip of the nozzle (which is usually of solid copper) by rubbing on fine emery paper placed on an absolutely flat surface, smoothing the surface of the jet tip until the face is once again absolutely level and true. There is a limit to the number of times that this can be done before a new nozzle is unavoidable.

An important point is correct replacement of the nozzle on the blowpipe. Two of the outer holes must be in line with the cutting jet so that a jet of heating gas precedes the cutting jet in order to ensure pre-heating of the metal. Which leads to the point that many novitiate operators tend to have the working pressure of the cutting oxygen far too high when cutting. This is a very great mistake, because too high a pressure can actually cool the metal to be cut.

Flame Cutting thin M/S sheet with a Step-Nozzle The step-nozzle is specifically designed for cutting thin mild steel sheet, which under a normal cutting blowpipe can simply melt or be burned away. In the case of the step-nozzle, part of one side is cut away and this area contains the orifice(s) for the heating flame. The cutting jet is incorporated in the full-depth position and this part is held down in close contact with the surface of the sheet metal, contrary to the normal cutting nozzle which is held some $5/_{16}$in. (8mm) above the surface. Care is needed in fitting a step-nozzle to make sure that the oxygen cutting jet is forward to the heating flame, although in the cutting process the flame precedes the cutting jet.

Flame-cutting relies on the metal being heated to incandescence before the oxygen jet can have any effect. In contrast to thin sheet, a large diameter round steel bar can be difficult to start. Holding a length of welding (or other mild steel) wire in the area of the flame against the bar will, as the wire becomes incandescent, promote the whole process and allow cutting to start immediately.

Flame Gouging. Gouging is a flamecutting process which can be used as an alternative method of weld preparation, when the normal methods of beveling cannot be used. Gouging is carried out by means of a specially shaped nozzle which cuts a groove in the workpiece to be welded, varying in depth as required. Needless to say, gouging is only carried out on very thick material, or perhaps to a defect deep in the material, to create a specially deep fillet, or even to remove a faulty weld.

Table for Oxy-Acetylene Flame-cutting.

Plate Thickness. inches	Nozzle Size inches	Operating Pressures. pounds/sq.in.	
		Oxygen	Acetylene
57	$1/32$	25	4
42	$3/64$	30	4
1	$1/16$	40	5
2	$1/16$	45	5
3	$1/16$	50	5
4	$5/64$	50	5
6	$3/32$	60	5

Remember – these tables are for HIGH PRESSURE BLOWPIPES **ONLY.** Low pressure blowpipes require different figures.

And may I add one final word of warning – never work without gloves and goggles, *and* a pair of tongs (or at least pliers) if one wants to avoid burns, which is the one hazard to constantly be on guard against.

Flame gouging

Fig. 24 Forming a groove by flame gouging.

Gas cutting and gouging

Recognition of cutting and gouging defects, their causes, prevention and permissible methods of rectification

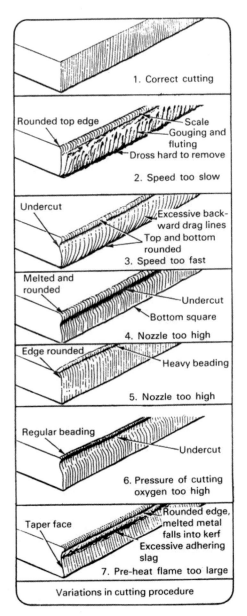

1. Correct cutting

Rounded top edge
Scale
Gouging and fluting
Dross hard to remove

2. Speed too slow

Undercut
Excessive backward drag lines
Top and bottom rounded

3. Speed too fast

Melted and rounded
Undercut
Bottom square

4. Nozzle too high

Edge rounded
Heavy beading

5. Nozzle too high

Regular beading
Undercut

6. Pressure of cutting oxygen too high

Taper face
Rounded edge, melted metal falls into kerf
Excessive adhering slag

7. Pre-heat flame too large

Variations in cutting procedure

1. In a correct cut the top of the cut is both sharp and clean, the drag lines are almost invisible, producing a smooth side. Oxide is easily removed, the cut is square and the bottom edge clean and sharply defined.

Drag lines should be vertical for profiles. A small amount of drag is allowed on straight cuts.

2. Due to melting, the top edge has become rounded. Gouging is pronounced at the bottom edge, which is also rough. Scale on the cut face is difficult to remove.

To rectify: traverse at recommended speed. Increase the oxygen pressure.

3. The top edge may not be sharp; there is a possibility of beading.

To rectify: slow down the traverse to the recommended speed. Leave the oxygen pressure as set.

4. Excessive rounding and melting of the top edge. Undercut has been caused by the oxygen stream opening out.
To rectify: adjust the stand-off distance between the nozzle and the plate.

5. Heavily beaded and rounded top edge, otherwise of good appearance.
To rectify: correct the stand-off distance by raising the nozzle to the recommended height.

6. The edge has a regular bead. The kerf is wider at the top with undercutting just beneath it.
To rectify: set the oxygen at the recommended pressure (on thinner steel it can cause a taper cut likely to give the impression that the cutter is set wrongly in relation to the plate).

7. Due to excessive heat, the pre-heat flame has caused the top edge to melt and become rounded. The kerf tapers from just below the top edge to the bottom of the cut face.
To rectify: set a pre-heat flame as recommended, use the correct nozzle at the recommended gas pressures.

43

PART TWO
Chapter 6

Arc Welding

Although we have already dealt with the question 'What is Welding?' perhaps it will not be out of place to re-define it here. Basically, it is simply 'the joining together of two similar metals by heating to a molten state, and, while in that state, fusing them together into one homogeneous whole'. Part 1 has already dealt with the method of achieving this by means of the gas-welding process.

Until about the end of the 19th century this was virtually the only satisfactory method available. It was excellent for the lighter jobs, such as light sheet metal and non-ferrous metals, and in the absence of any alternative, it perforce had also to be used for much heavier work, even for materials up to an inch thick or more. This was inevitably very slow and involved such heat diffusion that on really heavy work its use was greatly limited. About the turn of the century, engineers began to cast about for some alternative method, by which much heavier material could be welded, and naturally enough, turned to consider the possibilities of using electricity for this purpose. During their experiments with this, it was found that by passing a current of electricity at a

very high amperage, i.e. between around 50 to 200/300 amps, in conjunction with a correspondingly low voltage (approx. 45/60) an electric arc of high intensity could be caused to jump across a gap between the point of an electrode and the workpiece. The heat so produced was more than enough to melt mild steel, which (as stated earlier in part 1) melts at about 1450°C; the temperature of the electric arc is now known to be in the region of 6000°C. From then on this method was swiftly developed, at first simply by using as electrodes short lengths of bare soft wire, which, however, could only be used with D.C. current. The drawbacks of this were that the arc was difficult to maintain over welds of any length and needed very great skill on the part of the operator to keep it going. Subsequently, it was found that by giving the electrode a fairly thick coating of flux it could be used in conjunction with A.C. current, which gave a much more stable arc, and therefore of course, a much easier one to maintain. This is what opened up the enormous potentialities which we know today in the field of welding. With the most modern electrodes and welding machines and equipment now in use, there is practically no limit to the thicknesses which can be welded, leading to the ever-increasing use of welding in the world of engineering and construction.

Electric arc welding is universally used in engineering workshops and in industries like shipbuilding and ship repairing, as well as in the fabrication of large structures such as storage tanks. It is also used to a very great extent in car manufacturing for the welding of car bodies and components. Car bodies are all-welded into one whole, in what is known as a 'flash-welding' machine. The two sides, the roof, and back are inserted into the machine and tightly clamped, the edges being forced together by great pressure. A switch is pressed, there are a lot of flashes, and in seconds, the whole car body becomes one piece. This is known as

M.I.G. welding of small components in jigs. Note protective clothing – leather apron, gauntlets, strong boots – and safety aids such as fume extraction.

Notes on Safety – Electric Arc Welding

See that qualified electricians only are allowed to connect up power cables to the power source.

Always attach welding earth tightly to actual workpiece or bench.

Always avoid accidentally arcing wherever possible, and use fully insulated holders.

When holder is not in use, place it securely on an insulated hook.

Keep gas cylinders well away from arc welding activities.

Wear suitable clothing, including leather apron, and gloves or gauntlets and preferably rubber soled shoes or boots.

See that the hand screen, or helmet, is sound and has no holes in it.

Have adequate screens to protect other workpeople, and other welders, from the flashes of the electric-arc.

Goggles alone are not sufficient protection for the eyes.

If the operative does suffer from what is known as 'arc-eye' from excessive flashes, the eyes should be at once bathed with a good eye lotion.

Arc-eye is really conjunctivitis, and if it does not clear up within 48 hours then medical advice should be taken.

Never on any account have any loose connections anywhere, always see that all these are tight, with no arcing.

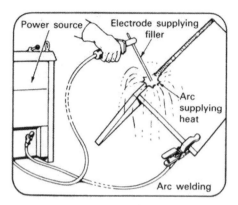

Fig. 25 *Arc welding principle.*

resistance welding (of which more anon).

There is a world of difference in places like motor car factories and properly equipped large engineering workshops where even very large fabrication work can be carried out (with the aid of hoists and lifting gear, enabling the workpiece to be turned round and over, thus enabling most of the welds to be done in the down-hand position, resulting in much neater welding

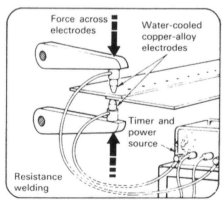

Fig. 26 *Resistance welding.*

beads) compared with a small village garage, a local smithy, or a model engineer's workshop, but the principles remain the same.

ARC-WELDING EQUIPMENT
(Machines and Accessories)

An arc-welding machine is basically simply an electrical transformer, which transforms, or converts, the electric mains supply from volts to amps, which is necessary for the purpose of arc welding, which requires a very high amperage. The welding transformer converts the high mains voltage (i.e. the usual two-phase power supply of approx. 440 volts) down to about 45/60 volts, with around 50 to 200/300 amps. There are smaller transformer sets aimed at the small user or D.I.Y. enthusiast which operate from normal single-phase 220-240v domestic supply; these are frequently advertised in hobby magazines etc. and can cope with most light jobs.

Some firms use their own electric generators, or rotary converters, but these deliver D.C. current, whereas the more commonly used transformer gives off A.C. current. Either form is quite suitable for arc-welding, and in fact in one case, i.e. for welding aluminum by electric-arc, using aluminum electrodes, D.C. current is essential.

However, for the purpose of this book, unless specifically stated it will be assumed that A.C. transformers are being used. Here again, these vary a little. Most of the bigger machines incorporate facilities for continuously variable current. These are far more advantageous, as with this type the operator can adjust the number of amps he requires for any particular job to within fine limits. With fixed-point types he is restricted to perhaps 5-10 points to plug into, which makes it impossible to obtain very fine adjustment of the current amperage; this is often a serious drawback.

Electrodes These are today almost universally of the coated type, that is, coated in varying thicknesses with flux. There are many different types but most are of the 'contact' or 'touch' type. With these it is possible to keep the electrode tip in continuous contact with the workpiece, which makes its very much easier for the operator to achieve a neat even edge. Electrodes come in seven standard wire gauge sizes, which with approximate metric equivalents are 16s.w.g. (1.6mm), 14s.w.g. (2.0mm), 12s.w.g. (2.5mm), 10s.w.g. (3.15mm), 8s.w.g. (4.0mm), 6s.w.g. (4.75mm), 4s.w.g. (6.0mm) and 2s.w.g. (7.1mm). Always keep electrodes dry.

Other equipment necessary is the electrode holder, which is attached by cable to the positive pole of the transformer, and the 'earthing' clamp, which runs from the negative pole of the machine. This clamp is attached to the workpiece itself or, as more often in general practice, permanently to the work-bench (which, of course, must be a metal one), or at least to a steel plate on top of the bench, on which the workpiece can be placed.

The welder's essential equipment includes a chipping hammer (with which to remove the coating of 'slag' after completing the weld; see later) and another essential is a wire brush, both for cleaning the surface to be welded and to clean up the weld afterward. A leather apron is advisable, and the same applies to a pair of leather gauntlets, again almost a 'must'.

The leather apron should be of the bib type with neck and waist straps, covering the whole of the front of the body. Asbestos-type gauntlets are an acceptable substitute for leather. The writer is aware that though these are compulsory under the Industrial Safety Regulations, many operators cannot be bothered with the leather apron because of its weight and the small amount of restriction of movement, but the writer urges habitual use of the apron for safety reasons. Splashes and blobs of molten metal can cause severe burns; even if you escape painful injuries, it can be ruinous to clothing. Beginners or trainees must cultivate the habit of wearing the regulation gloves and apron.

Screens These are of two types, hand screen and helmet with visor, and both incorporate special lenses. With the ordinary helmet type the operator has the advantage of having one hand free, to hold the workpiece while 'tacking' for instance, but it is sometimes an encumbrance when working in a cramped space (inside a small-sized tank for example), in which case the hand-screen may prove more convenient. For safety always inspect carefully that the screen, whichever type, contains no holes or cracks through which the intense ultra-violet rays emitted by the electric-arc can penetrate. These ultra-violet rays are extremely dangerous to the eyes, and exposure to them causes conjunctivitis, commonly known as arc-eye (inflammation of the eyes) as well as burning of the skin of the face. To any operator who values that most precious of all human senses, his sight, the only safe course is to scrap the faulty screen and obtain, if not a brand new one, at least one which is perfectly sound and gives full protection. I know it is tempting to merely patch a piece of sticky paper over the deficiencies, but believe me, this is useless as the ultraviolet rays can penetrate it easily.

The lenses are of varying opacity and should be fitted according to whether the job in hand involves heavy material, and consequently higher amperage, or lighter work.

SOME PRELIMINARY PRACTICE WELDS

Starting with the assumption that the operator is an absolute novice to whom the whole thing is a complete mystery, the first step is to switch on the machine, which we are assuming for the purpose of these first elementary steps is a simple ordinary welding transformer, already connected up to the mains by a qualified electrician.

For a few practice trial runs, any piece of scrap mild steel plate, say about ¼in. thick, will serve the purpose. The earthing clamp may be attached directly to this, or to the work-bench, (provided this is a metal one) or to a solid steel plate on the bench.

Now taking say (just for a first experimental try-out), a No. 12 s.w.g. electrode, and gripping the bare (uncoated) end of this in the electrode holder, set the welding current at about 140 amps. This is rather on the excessive side but will make things as easy as possible for the beginner to get started. Also have ready a small piece of scrap metal. Having donned gauntlets, take up the hand-screen in the left hand, if righthanded. (A left-handed person must reverse this, of course). Make sure that the odd piece of scrap material mentioned above is handy, and holding the screen well in front of the eyes (in arc welding one has to start completely blind at first until the arc is established, but this becomes quite easy with practice) 'strike the arc' on the piece of scrap. Striking the arc is, as much as anything, just like striking a match, and almost as easy. But do have the screen well in front of your eyes all the time. This is done prior to starting the weld proper, to gain two advantages (especially to the beginner). Striking the arc immediately before starting heats up the tip of the electrode and at the same time gets rid of any surplus flux or slag, or anything else which might obstruct the flow of the arc. Immediately after striking the arc on the piece of scrap, move over to the workpiece and while the electrode tip is still hot, strike the arc again, to commence your first welding bead on the workpiece. Now this may sound fairly easy, and in fact some novitiates have been known to start straight away and proceed to complete the whole bead of weld without a single break or stumble. But this is exceptional, and happens rarely.

What usually does happen, even with the utmost care, is that the beginner gets what is known as a 'dead short'. That is, on striking the arc, the novice fails to establish a sufficient gap between the tip of the electrode and the work, and consequently the dead short occurs, which results in the electrode welding itself to the work. When this happens, the unhappy novice tends to panic, and tries desperately to drag the electrode from the workpiece. This is absolutely the wrong thing to do, because in floundering about he usually lifts the complete workpiece off the bench, causing a lot of arcing all over the place, almost blinding himself temporarily, and anyone else who happens to be in the vicinity (and probably accompanied by urgent protests from the latter). All this is naturally rather frustrating and discouraging to the amateur, but there is no need for any of it, because there is a quite simple way of avoiding it. If you get a dead short, don't panic, simply release the holder from the electrode. Then, when cooled a little, it is a perfectly simple matter to break off the electrode from the workpiece. Thus no panic, and having avoided that feeling of frustration and annoyance you are ready to start again quite coolly for another try.

Although all this may sound a little formidable, it really becomes quite easy after practice, and may I repeat again, the secret of success is to strike the arc on the piece of scrap material first and then to 'get down to it' *at once* on the main workpiece, while the tip of the electrode is still hot. But above all, do keep the hand-screen (or helmet) in position shielding the eyes the whole time.

Practice Welding Beads. The operator, having surmounted these first hurdles and gained some dexterity and mastery of the technique, can now go on with the laying down of a few practice weld beads, or passes, as they are sometimes called.

It is important to establish a really good pool of molten metal right from the start; also, and this too is important, to get thoroughly used to being able to distinguish between the pure molten metal itself and the slag. In the process of welding, the flux on the electrode melts, along with the metal core of the electrode, and in doing so floats to the top of the pool of molten metal. To be more correct, this is what *should* take place automatically, providing the operator is skillful enough in his technique to ensure that it does happen. To attain this technique, the operator must become used to

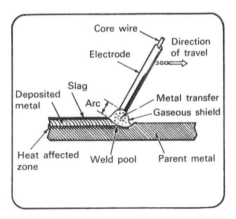

Fig. 27 *The welding arc.*

distinguishing this slag from the true metal. After some practice this is really quite easy, because the slag is a much brighter yellow than the pure molten metal, and easily distinguishable to the trained operator. The important point about all this is that the slag must never be allowed to infiltrate into the pool of true molten

Operator about to strike an arc

Fig. 28 *Ready to strike an arc.*

49

metal, as this impurity would weaken the weld. This is known as 'slag inclusion', and is something to be avoided at all costs, because it is fatal to good welding; it is the cardinal fault in the whole sphere of welding work.

The secret of ensuring that the slag does float to the surface of the weld is to keep the arc as short as possible during the whole process of depositing the weld metal. As the weld proceeds, the electrode burns, or melts, and the operator's object to maintain the melt at a steady and even rate along the whole length of the bead, and therefore, to do this, he must feed downwards at exactly the same rate as the electrode melts away. This demands quite considerable skill and finesse, but it is actually surprising how easy and natural it becomes after plenty of practice.

There are tricks in all trades, and here is one. Start the bead of weld right on to the extreme edge of the workpiece, thus getting the slag flowing over the edge, away from the pool of pure molten metal right from the start; it will then be found that it will almost automatically

continue to do so. Keep the arc as short as possible, along the whole length of the bead of weld, which greatly lessens the possibility of any slag getting into the pool of pure molten metal.

After the fledgling welder has completed his first run, and, after cooling, has chipped off the slag coating, the resulting bead of weld deposit should be revealed as an even, unbroken, uniform, and neat welding deposit, its surface resembling something like the following in appearance – ((((((((((((((– that is, a series of fine waves of weld metal. It is not expected that the beginner will produce this desirable result at his (or her) first attempt, but it is the ultimate object to be aimed at, and with reasonable practice, he (or she) should quite soon be able to achieve it.

One other point – after each welding bead is concluded, the slag has to be removed, but do not attempt to do this until the weld has cooled. To start 'chipping' immediately the weld is finished might result in a blob of still red-hot slag flying into one's eyes, with, as can be imagined, very painful and even dangerous effects. So make it an invariable rule never to chip the slag until the weld has cooled off. This has a double purpose, since the coating of slag protects the hot metal from atmospheric oxygen (which would otherwise cause oxidization of the weld metal) and also helps the weld to cool slowly, thus ensuring that it will remain soft, ductile, and therefore machinable – a very important property in a satisfactory weld.

When the operative has reached the point when he can consistently lay down a bead of weld metal neatly and evenly, and of equal width and penetration throughout, he can consider

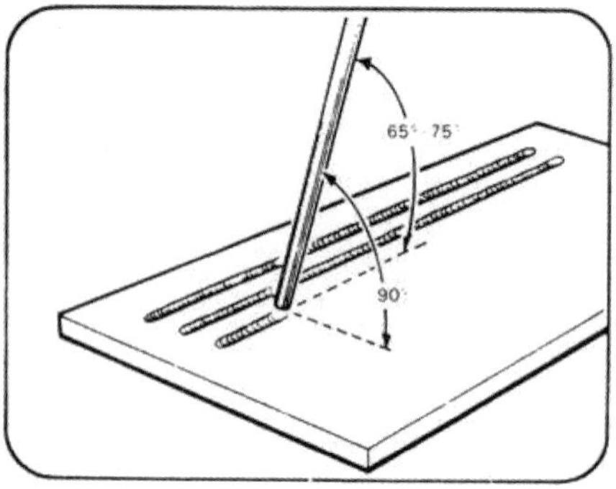

Fig. 29 *Correct electrode angle.*

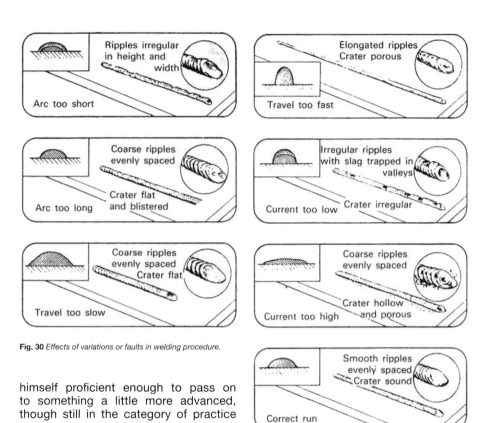

Fig. 30 *Effects of variations or faults in welding procedure.*

himself proficient enough to pass on to something a little more advanced, though still in the category of practice steps.

Chapter 7

Exercises in Welded Joints

THE SINGLE BUTT WELD

This in its simplest form is the union of two sections of mild steel of identical thickness into one homogeneous whole. (Other metals can, of course, be similarly joined, but these will be dealt with at a later stage).

As a practice butt weld then, let us take for a first attempt, say, two pieces of M/S plate of about $\frac{1}{8}$sin. thickness, and roughly each 6 x 3in. As a general rule, anything thicker than this would have to have the edges prepared by chamfering (or beveling), i.e. the edges to be joined must be ground to an angle of 30° on one side (in the case of a single butt weld). For material of not more than $\frac{1}{8}$in. thickness, edge preparation is not really essential, but for the purpose of a practice butt weld, we will do some edge preparation. For this thickness ($\frac{1}{8}$in., or 3mm) only one edge need be prepared to form what is known as a 'single-vee'. This is done by grinding each piece of plate along one edge, to an angle of about 30°, so that when lined up together, they form a total angle of roughly 60°. However, in grinding the chamfer do not grind this to a fine knife-edge. Always leave a light blunt edge or

'nose', the purpose of which is to leave a little solid metal on which to develop the pool of molten metal, without actually burning through. Also, in conjunction with this slightly thick edge, we always leave a slight gap between the two sections to be welded. So, in lining up our two practice pieces of M/S plate, leave a slight gap of approximately $\frac{1}{16}$in. (1.5mm) at the bottom of the vee. Place the plates in alignment horizontally on the earthed bench and clamp them down so that the operator can work from left to right, if a right-handed person. A left-handed person would, of course, have to reverse this.

We will, for this thickness, require a No. 12 s.w.g. electrode, the end of which is now bared and fixed into the holder. The welding current should be set at approximately 120 amps. This is slightly less than that suggested for the practice runs, although our material is of the same thickness. This is because we have now prepared the edges by grinding, thus reducing the thickness at this point. With the screen in position, 'strike the arc' as before explained, and immediately proceed to start the weld,

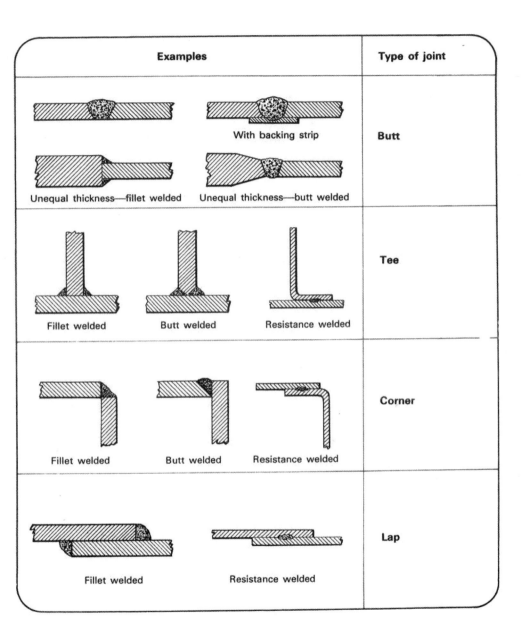

Fig. 31 *The four basic forms of weld applied to the commonest types of joint.*

commencing at the end opposite the tack. Beginning well over the left-hand edge, to get the molten slag flowing well away from the pool of pure molten metal, steadily proceed from left to right, steadily feeding down the electrode at an even rate as it melts away, and maintaining the width of the welding bead, so that it embraces the whole width of the vee, along the whole length of the seam. In carrying out the above practice weld, two points should be borne in mind. First, the operator must try to achieve 100% penetration throughout the whole length of the seam, and he must practice this until he can complete the weld without the slightest speck of any 'slag inclusion' whatever. After the weld is completed and cooled, it should be possible to saw through at any point, with the cut edge showing nothing but 100% solid metal right through. Second, after completing the weld on the 'vee' side, it is advisable to turn the practice piece over, and place just a slight 'sealing run' on the reverse side, using the same size electrode and current setting. Strictly speaking, this should not really be necessary because, theoretically, complete penetration should have been achieved on the first run on the veed side; this is what the vee is for. However, it is unlikely that the novice will be able, in fact, to achieve this desirable result in his first few attempts, so the 'sealing run' on the reverse side is a wise precaution. Also, there is another point: quite possibly (probably, in fact) the M/S plate on cooling may have warped, i.e. become bent upwards. This is caused by the contraction of the main weld (on the vee side) on cooling. A sealing run, or bead of weld, on the reverse side will help to correct this. However, more on

the problem of distortion will be dealt with later in the book.

DOUBLE VEE BUTT JOINT

For M/S of ¼in. (6mm) and over, it is advantageous, wherever possible, to use a double vee edge preparation. I emphasize 'wherever possible', because in many cases it is just not practicable for various reasons to weld both sides of a job. As an example, the joint may have to be welded after the work has already been placed in position and the underside becomes inaccessible. However, for our present purpose which is purely that of a practice double vee butt weld, the plates can be easily assembled on the bench surface.

Using two plates similar in size to those in the previous exercise, chamfer the edges to be welded by grinding an angle of 30° at both top and bottom, leaving an unground strip along the center line of each edge. When the two pieces are placed on the bench they will show an angle of approximately 60° at both top and bottom, but they should be separated by a gap of about $^3/_{32}$in. (2.5mm) along the center flats. Clamping to the bench both holds them securely and minimizes the effects of distortion.

A tack-weld should be made at each end of the seam. These tacks are simply small blobs of weld metal placed at strategic points, their main function being to hold the separate parts in correct relationship while the main weld is made. It is important that each tack is a proper weld, i.e. true fusion of the components is achieved, because it often happens that a whole, heavy fabrication may be held by tacks until the assembly is advanced enough for the actual welding procedure to begin. Every tack-weld should be a true fusion weld and long enough to hold parts safely together. Many otherwise proficient operators do not fully appreciate that getting into the habit of making sound tack-welds can on occasion prevent risk to life and limb.

To return to our practice double vee butt-weld, having tacked the two component pieces together while clamped to the bench, the workpiece can now be released from the clamps and turned over, so that the initial welding run can be carried out on the opposite side. The idea of doing this is to minimize distortion, which is an ever-present danger, a major cause of which is the contraction of the weld on cooling. To overcome this the aim where possible is to counteract each welding run (or bead) by placing an opposing bead of weld deposit on the reverse side, thus causing the forces of expansion and the accompanying contraction to work *for us* instead of *against us*. It is hoped that the reader will take particularly special note of the above passage, because it is a vital point which contains a whole world of significance, and because this simple strategy contains the secret of the means by which that great bug-bear of welding, distortion, can be defeated.

Having turned the workpiece over, the weld can now be carried out in the vee on the reverse side to that of the tacks in the normal way. On completion of this bead, the workpiece is again turned over and the final weld carried out on the tacked side. This is known as a setting run, or setting bead. After cooling the slag can be chipped off (a lot of it will probably have already fallen off), and the completed workpiece should now present one whole section of M/S plate. If desired, the surplus deposit can now be

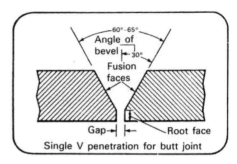

Fig. 32 *Single vee edge preparation.*

ground off, which should then render the weld completely invisible, with no sign of a joint whatever. If you wish, the plate can (as a test) now be sawn through at any point across the line of weld, and on inspection, should show nothing but solid metal throughout.

SINGLE AND DOUBLE VEE BUTT WELDS IN HEAVY M/S FABRICATION WORK

It is in the field of heavy plate and angle work that electric-arc welding really comes into its own, because there is almost no limit to its possibilities, or to the thicknesses which can be welded by this process. Also the speed by which work can be executed far exceeds that of oxy-acetylene welding – it is in fact considered to be up to six times as fast. However, each system has its own sphere, and the two processes do, in fact, complement each other. Oxy-gas welding is supreme in the field of light sheet metal and non-ferrous metals such as aluminum (though this is now welded by a combination of the two, known as the shielded-arc method, of which more anon) while arc-welding is superior in the sphere of heavy

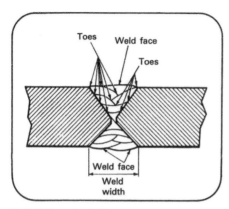

Fig. 33 *Multi-runs in a double vee.*

engineering. Frequently both processes are used: heavy M/S plates and angles are cut by oxy-acetylene, preparatory to being arc-welded. In large engineering works, the cutting-out is often carried out on automatic profiling machines, of which there are various kinds; there are also oxy-acetylene planing machines, which are used for the purpose of chamfering the edges of the work to

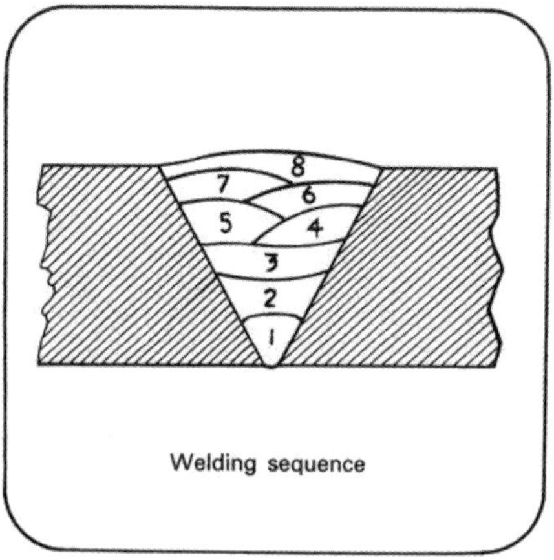

Fig. 34 *Order of applying multi-runs.*

be arc-welded later. I have re-referred to the above in order to illustrate how the two systems can complement each other in economical and commercially viable techniques.

To revert, however, to our original theme, that is, single and double-vee butt welds on heavier material. In these cases, because the vee has, of necessity, to be much deeper and wider, more welding beads, or passes, will be found to be required. These are called 'multi-runs', and must be placed in their correct order or sequence. First, the initial run or 'root' bead is carried out in the normal way, (on both sides in the case of the double vee), and then the following beads are laid in the sequence illustrated. For the benefit of the beginner, there is an important point. After each run or bead the slag must be completely removed before starting each succeeding bead of weld deposit. This is absolutely essential, to make completely certain that there is no possibility of the slightest speck of slag inclusion in the weld.

FILLET WELDS

This joint is perhaps the most frequently and widely used of all the welded joints, constantly needed in all engineering and construction work. It is also, at the same time, probably the easiest to carry out. In its simplest form, it merely involves the running of a bead of weld deposit along the inside of the juncture of the two sections to be welded, which are set at an angle of 90° to each other.

A Practice Fillet Weld on the bench.
For our purposes, two similar pieces of M/S plate to those used in the previous exercise can be used, i.e. two sheets of

56

about 6 x 3 x 1/4in., but these must now be set standing in an upright position on the bench, at an angle (roughly for this simple exercise) of 90° to each other. There are specially designed corner clamps sold for this purpose, including magnetic clamps, which are very useful indeed, and if any of these are available, they make the job of holding the two parts in position until a tack can be placed very much easier. However, in the absence of a special corner clamp, the two sections may be stood on end on the bench at the 90° angle, and if the operator is wearing a helmet, he can hold the two parts in position with his free hand while tacking. If only a hand-screen is available, and no-one by to help, a tack may be placed on the top edge of the corner. However, here again we fall foul of our old enemy distortion, because as soon as the tack-weld has cooled, its consequent contraction will cause the two sections to open apart at the base (or bottom ends) of our practice pieces. There is, as it happens, a quite simple way of getting over this problem which can be classed as another 'trick of the trade', but the operator must be fairly 'quick on the draw' to carry it out. This is how it is done. As soon as the operative has placed the tack, he drops the screen and grabs the base of the workpiece before it has time to cool off, holding it while the tack is still hot. Sounds simple enough, doesn't it? It *is* quite easy, after a little practice. The workpiece can now be laid horizontally on the bench, when it is perfectly easy to place a corresponding tack on the opposite end.

So, now with our practice piece lying on the bench open side up propped by any odd piece of angle or other material, it is now in the perfect position for our practice fillet weld. The procedure for this is very little different from that of the single-vee butt weld – we have a similar vee in which to place the bead of weld. A similar size electrode (12 s.w.g.) is used but the current needs to be set a little higher than that for a single vee, because a fillet weld requires a little more heat in order to achieve sufficient penetration. The procedure is as before, working from left to right, striking the arc, establishing a good pool of molten metal, and keeping the arc as short as possible. Proceed steadily along the root of the vee, feeding the electrode down at a steady rate as it moves along, and all the time keeping a close watch to make sure that there are no slag inclusions. The reader will have gathered by now, I am sure, that arc-welding demands intense concentration throughout the whole duration of the welding operation, and the utmost vigilance to guard against the danger of slag inclusion, but after reasonable experience this becomes automatic.

Distortion, as referred to earlier, is the one other big problem which constantly confronts the operative in the sphere of welding, but there are ways and means of overcoming this, as witness the examples cited earlier. All it really needs is the exercise of a little thought and strategic care in the order in which the successive beads of weld deposit are placed. In the case of very large and bulky sections of plate, other methods than simply laying beads on opposite sides have to be used to combat this fundamental and ever-present problem. Devices such as wedges, combined with cooling systems, and a special sequence in the order of laying the welding beads (known as 'backstepping') all have their place, but the

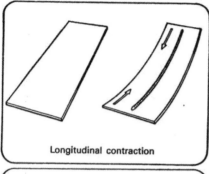

Longitudinal contraction

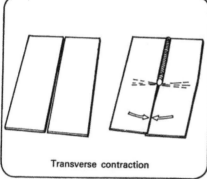

Transverse contraction

Fig. 35 *Contraction caused by the weld bead.*

Fig. 36 *Balanced welding.*

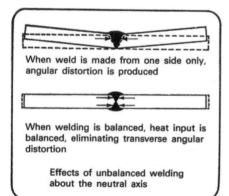

When weld is made from one side only, angular distortion is produced

When welding is balanced, heat input is balanced, eliminating transverse angular distortion

Effects of unbalanced welding about the neutral axis

novitiate welder need not concern himself with these at this present stage. Sufficient to reiterate that all these stratagems are based on the principle of making the forces of contraction and expansion work for us instead of against us. This strategy can be very effectively used in the fillet weld which we are now discussing.

On completion of the first bead of this, it is inevitable that contraction will take place on cooling; the effect of this contraction will be that the two 'flanges' or sides of the angled plates will tend to close inwards toward each other, on the same side as that on which the welding bead has been placed. Knowing in advance the direction in which the integral parts of a welded fabrication are going to move enables

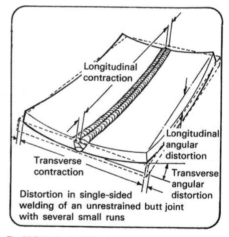

Distortion in single-sided welding of an unrestrained butt joint with several small runs

Fig. 37 *Distortion in all directions.*

us to defeat the forces of contraction. The key is that the direction in which contraction takes place after cooling is always to converge *toward* the bead of weld which has just been deposited. This was seen in the simple practice single-vee butt weld, where on completion the outer edges of the plates tended to rise upwards from the bench surface. If the workpiece had previously been clamped to the bench, or some other firm solid surface, then this effect would have been lessened, but whatever contraction and consequent distortion had taken place was counteracted by the 'sealing' run which, it may be remembered, was placed on the reverse side. The use of a second bead of weld deposit on the opposite side to the first is in fact the most effective and simplest counter to distortion, where it is possible or practicable to lay one immediately after completion of the first; in this way the use of clamps, jigs or special cooling arrangements can be avoided or, at least, the need for them reduced.

Applying the principle to our fillet weld, the first bead has caused the sides of the vee to close inwards and if a second bead is now laid on the outside of the joint, the anticipated contraction should restore the vee angle to its original 90°. This we can proceed to do by simply turning the workpiece over and laying a bead of weld on the side now uppermost. Incidentally, if, in setting up the two sections (before welding) we placed them edge to edge, this should now present a perfect vee in which to lay down the second bead of weld. In effect it forms a 'corner weld', which we now come to.

CORNER WELDS ON MILD STEEL

The two abutting edges of the plates above, placed corner to corner, clearly produce a 90° vee angle in which to lay the weld deposit. In this case, where we are only using ¼in. (6mm) M/S plate, one 'run' or bead of weld will be found to be sufficient to build up the corner to the level of the surrounding material

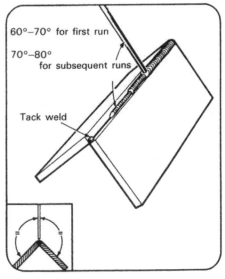

Fig. 38 *Corner joint.*

Fig. 39 *Avoid reducing plate section.*

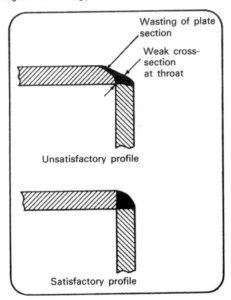

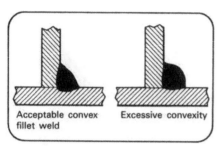

Acceptable convex fillet weld   Excessive convexity

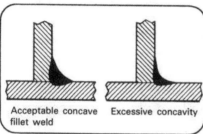

Acceptable concave fillet weld   Excessive concavity

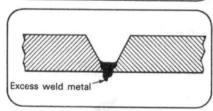

Excess weld metal

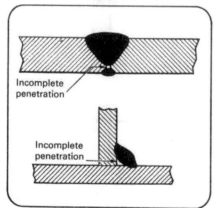

Incomplete penetration

Incomplete penetration

as illustrated, and here too, is an important point – the level of weld deposit must be built up to the full strength of the adjacent steel plate section, just as if, in fact, the plate had simply been *bent* round to the 90° angle. When the surplus weld material has been ground off, the workpiece should present just that appearance.

Before leaving the subject of fillet welds, a word or two should be include on the two main pitfalls to be avoided.

1. 'Undercutting'. This, briefly, happens where too high a current has been used, causing the electric-arc to burn into the material at the edges of the weld bead and creating a gouging effect at these points, which is very bad practice because it obviously means that the metal is greatly weakened. This then is a danger to be avoided, and to do so, beware of setting the welding current too high. The effects of undercutting are clearly shown in Fig. 41.

2. The other danger to be wary of is the very opposite to the above. This is *insufficient penetration* due to using too low a welding current. This can be as serious a defect as that of undercutting, though in a different way. Because the welding current is too low, the arc cannot penetrate deeply enough into the steel to achieve a really strong joint. The sketch illustrates this very essential consideration in all welding work, which cannot be too greatly emphasized. The welding current must be set high enough to achieve deep penetration of the material being welded, while avoiding the corresponding danger of having the current so high as to cause weakening of the material by undercutting at the edges of

Fig. 40 *Faults to avoid.*

the weld. These factors are particularly applicable to fillet welds, but nevertheless, the same principle is of equal importance in butt welds and, in fact, in all welded joints.

FILLET WELDS ON HEAVIER MATERIAL

Obviously, when very much heavier and greatly thicker material has to be welded, one welding bead would be totally insufficient and several beads (or passes) have therefore to be placed at the joint. These are called 'multi-run fillets' and they must be placed in their correct sequence as shown. After the first or root run is completed, the second run, or bead, should be placed at the base of the first, slightly overlapping it. The third run is now placed above the second, again overlapping that. See Fig. 34. The reason for this procedure is to ensure that the whole of the multi-run fillet is one homogeneous whole, i.e. solid metal throughout. If the beads of weld deposit were placed side by side, a weak spot would be left between each succeeding bead, an inadmissible fault which could cause a serious fracture.

'T' JOINTS

These are, in effect, very similar to the ordinary fillet welds, except that the vertical plate (the stem of the T) is placed on the horizontal one to leave what might be termed a flange requiring welding each side.

The technique and procedure are much the same as in the fillet weld, but here again, be it noted, we are still up against the same old enemy, distortion. However, and to reiterate, by using the technique of opposing each bead of weld by an equal bead on the opposite side, this problem can be beaten.

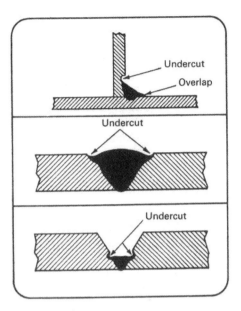

Fig. 41 *Undercutting.*

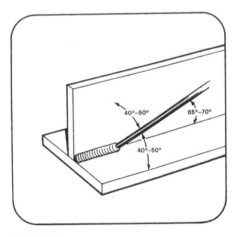

Fig. 42 *T-joint.*

61

A Practice Weld on a 'T' Joint. Briefly, the procedure is as follows. First, set up the two practice pieces (similar plates to those used for the practice fillets will be excellent) on the bench at a 90° angle for tacking, held in place by the aid of a magnetic clamp if possible. If this is not available, then the 'grab' dodge can be used as described earlier. Incidentally, this is a case where the helmet type screen comes into its own, because as it leaves the operator with one hand free, he can hold the two sections together while he places the first tack on the top edges of the workpiece, and also, at the same time, he can hold against the forces of contraction until the tack has cooled off, thus comprising a double advantage.

The practice workpiece can now be placed horizontally on the bench, and as was done in the practice fillet, supported by a piece of packing in the vee position, and the welder can now proceed exactly in the same manner as in the practice fillet, taking care, as emphasized above, to place a bead of weld on each side alternately, in order to counteract the distortion effect. If carefully followed, this procedure should result in a perfect 90° angled tee joint. When completely cooled off, a check can be made with a steel square.

The reader may perhaps wonder why I always advocate laying down the practice piece on the bench in the horizontal position. The reason is that it is far easier to weld in what is known as the 'downhand' position, i.e. horizontally. Of course, in fully operational welding work it is very often necessary to carry out a weld in the vertical position, and also very often overhead. This is moving into the province of the expert professional welder, however, and although some notes on such welds are included later, where possible downhand techniques are recommended for the amateur.

On very heavy sections of material where a tee joint is required a slightly different method of edge preparation has to be used. There are two approaches, the single or double 'U', and the single or double 'J'.

THE SINGLE AND DOUBLE 'U' JOINT

When a weld is called for in exceptionally heavy material, say from about one inch thickness upwards, the edge preparation required for the vee would have to be of such enormous proportions, and would entail such a vast amount of grinding or cutting, or other means of chamfering, with the consequent time involved, that it would not be economically viable. However, there is a simple alternative method which can be employed for the purpose of preparing the edges to be butt-welded. This is known as a 'U' joint. When only one side of the material is accessible only a single 'U' is possible, but if both sides are workable, then the double 'U' edge preparation should be employed wherever possible.

There is a very easy way of forming these kind of edge preparations which is not only simple and effective, but involves a minimum of time and very little effort. It merely consists of laying a bead, (or two or more beads) of weld deposit, using a fairly large size electrode (say number 8 s.w.g.), along the

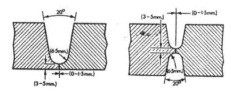

Fig. 43 *Single and double U-joints.*

Fig. 44 *U edge preparation.*

bases of the edges to be joined, on both sections, which, when placed together then form the desired single 'U'. The same procedure can be utilized for the double 'U', except that in this case, the bead (or beads) of weld metal should be placed in the center of the edges of the sections to be joined as shown in Fig. 44.

A single or double J joint is similar in concept, but applied to tee joints rather than butts.

Important Note In welded joints of this kind, many more than one welding bead or pass will be found necessary, and the order of procedure is the same as that obtaining in the multi-run fillet welds, that is, in the same correct sequence. It is worth repeating that the coating of slag must be completely removed after the completion of each successive bead before commencing the following one. It is impossible to stress too highly the vital importance of this. Also, in the case of the double U or J joint, the succeeding beads of weld deposit should be placed alternately on opposing sides, in order to combat the effects of distortion as described earlier.

Further, if it is at all possible to do so, with a single U or J, a sealing run on the

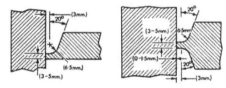

Fig. 45 *Single and double J joints.*

underside is advisable. If this is not possible, then great care should be exercised in laying the first or root run, ensuring that this really does penetrate right through the material, even protruding slightly through the underside at the base.

LAP JOINTS

As seen in the illustrations, the lap joint really consists of two fillet welds, one at each end of the sections to be welded. Whether these can be done with a single bead, or pass, of weld deposit depends on the thickness of the material. Thicknesses of up to about $^3/_8$in. (9mm) can be successfully welded with a single bead using a No. 8 s.w.g. electrode, with the appropriate amperage, but for anything thicker than that, two or more runs will be found necessary.

A Practice Lap Joint. For this two more pieces of M/S plate of 6 x 3 x ¼in. can be used, but in a lap weld joint, the two sections are clamped together, one on top of the other and overlapping, leaving a space at each end of the weld. For this thickness a No. 10 electrode will be found sufficient to complete the weld with one bead. After firmly clamping the two sections together, the next step is to place a strong tack at *both the upper and lower* sides of the practice piece. This is essential because, even with the clamps on, a certain amount of distortion can take place, especially if the clamps were mistakenly removed before the final welds are carried out. So the moral is, tack very strongly at each side before removing the clamps. (In an actual welding construction job, the clamps may *have* to be removed prior to welding).

63

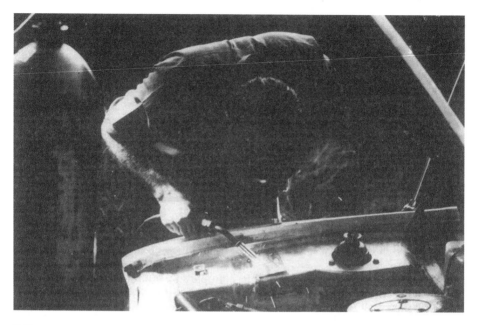

Welding a patch under a car bonnet is a common repair and is effectively a fillet joint round all four sides of the patch. Note earth clip, center foreground.

Now, with the welding test piece lying flat on the bench, commence to weld (it doesn't matter which side first), starting at the end opposite to the tack: in other words, so that the bead of weld finishes on the tack. Incidentally, here is a maxim which in fact applies to all welding. Whenever a welding bead is laid down, the heat builds up progressively as the weld proceeds. Therefore, as the end of the run is reached, the heat is built to a much greater maximum to that at the beginning, so, unless the operator is careful to avoid it, the edge of the material at the conclusion of the weld may become burned away, leaving a large cavity or crater. This would constitute a very serious weakness in the joint and to avoid such a problem the experienced welder always *finishes* on the tack, or,

alternatively, builds up the weld at this point with extra weld deposit to compensate.

There is also one other important particular concerning lap welds. This is to make sure that the bead of weld deposit completely fills the vee, or fillet, formed at each end of the workpiece, i.e. the bead of weld metal should reach right to the top surface. If the metal does not fill the space but leaves a gap unwelded, the danger is that this again leaves a weak area, and the material could therefore split at this point. The

Fig. 46 *Fillet-welded lap joint.*

64

golden rule is to make sure that the weld fills up the whole of the fillet or vee – any surplus material can be easily ground off later if necessary.

Arc-welding of Very Thin Mild Steel Sheet

Let me say at the outset that thin mild steel sheet (say 16 s.w.g. or less) is far better welded by gas (oxy-acetylene) which results in a far superior weld. It is extremely difficult to weld by the electric-arc, which simply burns right through the material, or the amperage has to be cut so low that it is very difficult to get the electrode to run. In fact, many otherwise highly skilled welders cannot do it at all. However, if no gas welding plant is available, then sheet metal work will perforce have to be welded by the electric arc.

The best way to do this is by using a 'chill block' consisting of fairly thick slab or bar of steel, not less than ¼in. thick. If copper is available, so much the better, because it is such a good conductor of heat – which, as no doubt the reader will quickly guess, is the whole idea.

The pieces of sheet metal to be joined are tightly clamped to the block at both ends by G-clamps and a No. 16 gauge (1.6mm) electrode used, with about 50 or 60 amps. One run will be sufficient, but if necessary, a light sealing run can

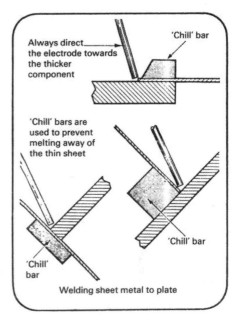

Fig. 47 *Some chill-bar applications.*

be deposited on the reverse side. To make it easier the whole assembly, sheet metal and chill-bar, can be set up at an inclined angle (of about 30°) and welded in a downwards direction, at a fairly fast speed, but do not attempt to weave. For fillet welds and corner joints the same technique may be used, with the chill bar clamped accordingly, as shown in Fig. 47.

Chapter 8

Cast Iron and Stainless Steel

There are three kinds of cast iron in general use, the two commonest being white cast and gray cast. There is also malleable cast iron.

Of these, gray cast iron is the most easily weldable, as it is far less brittle than white cast iron, because of its lower carbon content and greater proportion of silicon. Gray cast is also far more easily machinable. White cast iron is not regarded very favorably for welding purposes at all.

Malleable cast iron, as its name implies, is far less brittle than either gray or white cast iron. Its peculiar quality is brought about by a process of annealing, involving a very slow prolonged period of gradual heating to a high temperature, followed by an equally prolonged period of slow cooling. Sometimes the casting is buried in quicklime for this purpose for several days, or perhaps even weeks, in order to achieve the degree of ductility or malleability required. This is a long and costly process, and even then only achieves a soft outer skin, of greater or less depth, while the center still remains relatively hard and brittle.

It is the last feature which causes malleable cast iron to present special problems in welding, because, in effect we are dealing with two distinct metals in one, the soft outer skin, and the inner hard core of metal. For the purpose of this book discussion will be confined to that of welding *gray* cast iron, though malleable castings can be dealt with by brazing (or bronze welding) which has already been dealt with earlier.

In considering the welding of cast iron the perennial question is always which method of welding is preferable, gas or electric-arc? This, like so many other problems, may finally be decided by expediency. No gas welding plant may be available, or the job may be far too large and massive to use the gas-welding method with no pre-heating furnace available.

If a pre-heating furnace is available, and the job is not too large and bulky, then gas-welding probably does hold the advantage, because when all the factors are considered, this process on the whole produces a more satisfactory job from the point of view of machinability and general soundness of the weld. Also, with the gas-welding method, the filling material, (ferrosilicon or similar filler rods) has a greater affinity with the parent (cast iron) metal than that used in the arc welding method, the filler metal of which is nickel alloy, or similar. Having said that, however, the arguments in favor of the arc-welding of cast iron rest on the grounds of speed, convenience, accessibility, and general circumstances, as well as economic considerations.

A further point in favor of the electric-arc is that the heat is kept far more localized and the heat-spread restricted, therefore greatly reducing the danger of cracking of the casting, though clearly in considering this a great deal depends on the shape of the casting. If, for instance, this is just a simple straightforward cast iron bar, where the ends are not secured in any way and are therefore completely free to expand and contract, then welding it is a comparatively simple matter presenting no difficulties. No pre-heating is necessary, and it can therefore be welded in the routine way, by either gas or arc-welding technique. If on the other hand the casting to be welded is not so straightforward, or in cases where the ends are 'locked up' (as for instance a fractured spoke of a cast iron wheel), then pre-heating of the whole casting would be essential, and for this purpose a muffle furnace big enough to take the job is necessary. If a proper purpose-built muffle furnace is not available, it might be possible to build a temporary makeshift one round the job.

Preparation for Cast Iron Welded Joints (Arc-Welding). Edge-preparation for arc-welded joints in cast iron, i.e. for butt welds, fillet welds, tee joints, corner welds, etc. is exactly the same as that for mild steel, and the procedure is almost identical, except that the elec-

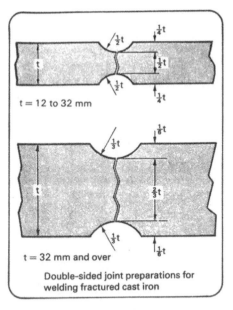

t = 12 to 32 mm

t = 32 mm and over

Double-sided joint preparations for welding fractured cast iron

Fig. 48 *Preparation for cast iron repair.*

trodes used are of the nickel alloy type. These are quite easy to use, with the arc always kept as short as possible. Also, after each run as with mild steel, the slag must be completely removed before commencing each following run.

A Simple Exercise in a Double-Vee Butt Weld of Cast Iron. For this purpose let us take just a simple straightforward piece of casting – say, for example, the leg of a turning lathe, or anything of similarly straight shape, which has broken in a clean fracture, leaving the ends fairly square, and of about say, 3 x ¾in. in thickness. A casting of this kind, being completely free at the ends, and therefore involving no compression problems on expansion, as explained earlier, will not need pre-heating, although no harm is done if gas equipment is available if the flame is played over the two sections as a preliminary before actually starting arc welding. This would remove the chill from the metal and the warming-up effect helps to save time, especially on a really massive casting. In addition it makes starting easier, but it is not absolutely necessary.

Edge preparation for a double-vee butt-weld on cast iron is the same as that for mild steel, i.e. chamfering by grinding to 30° angles on both sides, leaving, however, a slight section of the casting untouched at the center. This is particularly useful if a broken casting is being repaired, since leaving a little of the fractured edge of the casting untouched by grinding allows the broken edges to line themselves up beautifully ready for chamfering in position.

Using a No. 10s.w.g. nickel alloy electrode, set the current at about 140 amps, bearing in mind that cast iron melts at a much lower temperature than mild steel: gray cast melts at 1240°C, while mild steel melts at about 1430°C.

The actual technique of welding cast iron by the electric arc is not very different from that of mild steel. There is one slight difference – the flux coating of the nickel alloy electrode is very much lighter, resulting in not quite so much slag to be removed, but again it must be stressed that every particle must be removed after each bead of weld deposit before commencing a following one. The arc must be kept as short as possible throughout. Assuming that the two sections are perfectly lined up, first place a strong tack at the edge on one side, to hold the pieces in position, then turn the workpiece over and proceed to weld in the normal way as with mild steel, from right to left. After completing one bead of

weld deposit on this side the workpiece should be turned over and the second weld proceeded with on the other (tacked) side. Once again, we are balancing two beads of weld to combat distortion. On completion of the weld on both sides of the double-vee, the casting on cooling should finish up perfectly straight and in accurate alignment. With cast iron, cooling should be gradual – as long and as slowly as practicable, because sudden cooling could cause cracking, as well as rendering the weld deposit too hard for subsequent machining. To help slow cooling of the casting, it can be covered in dry sand or by some other protective material to exclude as far as possible cold air and drafts. If this procedure is followed the resulting welded joint should be 100% sound and quite ductile and machinable.

Fillet Welds and Tee Joints in Cast Iron. It will be found that these types of joints are not very frequently met with in castings, in general engineering work, probably because most castings are gusseted or otherwise reinforced at such junctions, and are unlikely to break at these parts of the section.

If, however, the occasion does arise and such a weld becomes necessary on a casting, the welding procedure to be followed conforms exactly with that for mild steel fillet welds and tee joints.

ARC-WELDING OF STAINLESS STEEL

Stainless steel in reasonably thick or heavy sections lends itself very well to welding by the electric-arc process, using stainless steel electrodes intended for the purpose. Stainless steel sheet is better welded by the gas-welding method, described earlier, but best of all by the T.I.G. or M.I.G. processes to be described later. However, it must be remembered that there are today many varied kinds of stainless steel and care must be taken to select the correct and most suitable electrode for each particular type.

Apart from this important point arc welding of stainless steel does not differ greatly from the normal technique used with mild steel. There is a slight danger sometimes of a tendency for the completed weld deposit to crack on cooling, but this 'hot cracking' as it is called is usually due to the weld having been carried out too fast, with too high a current. If this trouble persists, a way of obviating the danger is to lay a bead of ordinary mild steel weld deposit on to each face of the chamfer of the vee prior to proceeding to weld with the stainless steel electrode. This is known as a 'buttering layer' and it helps to prevent any cracking of the final beads of stainless steel electrodes. It should not be necessary to stress once again that it is vital that every particle of slag must be got rid of after each bead is laid, before proceeding with the following one.

Edge Preparation. Edge preparation of stainless steel for butt welds, both single and double, and for tee, single and double 'U' and lap joints is exactly the same as that for mild steel.

Chapter 9

Pipe Welding

Pipe (or tube) welding is really a separate and distinct branch of its own in the sphere of welded production work and those welders who do engage in this class of work are usually specialists, in the sense that they have undergone a period of special training and experience to acquire specialized skill. Such skill is very necessary for pipe welding, because pipes and other tubular constructions, by the very nature of their shape in the round, present their own peculiar and unique problems in jointing and fitting. Pipe welding most frequently arises with heating installations in building and construction work and – especially in recent years – in connection with gas and oil pipelines. The latter two are often of very large diameter, frequently of two feet or more, with the thicknesses of the pipe walls in proportion.

Another special feature of pipe welding is that many of the joints have to be carried out after the pipe is placed in position – perhaps at the bottom of a deep trench for instance, with perhaps just enough room for the welding operator to crawl underneath to carry out his task. A further point is that because of their shape, pipe and tube welding involves a lot of positional work, i.e. it includes vertical and overhead welding as well as the normal downhand (horizontal) welding positions. Additionally, in the case especially of gas or oil pipework, the joints will on completion have to be subjected to stringent tests, such as examination by X-ray, and proved able to withstand pressures of up to 1,000lbs. per square inch or more. It will readily be seen from this why pipe welding is such a specialized branch of welding. However, for the sake of completeness some descriptive exposition of this particular work should be included, so let us review the more usual and normal welding joints as they are applied to pipework.

Edge Preparation for a Butt Welded Joint on (for example) a 6in. dia. Pipe or Tube. The edge preparation for this is, in its essentials, the same as that for a single vee butt weld on flat mild steel plate, except of course in this case we are dealing with a circular-shaped workpiece.

First we have to get the two sections into perfect alignment and a simple way to accomplish this is to form a cradle by using a fairly long section of large angle-iron, fixed in the vee position. Alternatives would be rollers or a length of iron channel large enough to allow the two sections of pipe to be placed inside and clamped. They need to be butted close together but with a small gap of about $1/_{16}$in. or 55in. according to thickness of material. The two sections of pipe can now be tacked at about three or four places around the pipe, after which the weld can be proceeded with in the normal way, in the downhand position, turning the pipe round as the weld progresses to suit the operator's convenience, but keeping it firmly clamped while actually welding. Great care must be taken when changing electrodes to commence a fresh pass, which must maintain perfect union with the previous one, i.e. there must be no gap or break in the continuity of the weld, no matter how many electrodes may have to be used. The finished weld, on completion, should present one unbroken whole bead of weld deposit. Obviously, on much larger diameter pipes, one bead of weld will not be sufficient to completely fill the vee, even with larger size electrodes, and several multi-runs may be necessary. For the finishing run it may be found necessary to weave slightly, i.e. use a side to side motion of the electrode while welding proceeds, in order to ensure that the final bead of weld metal covers the whole width of the vee and to make sure of complete fusion of the edges of the weld with the parent metal.

The above is just a simple elementary exercise for a butt weld on a comparatively small size pipe in which the operator can turn the pipe round to suit his convenience, but it should be borne in mind that in large-scale constructional work, such as gas or oil pipelines, this is rarely possible. The welded joint must usually be made round the stationary pipe, which involves quite a lot of vertical and overhead work, often in extremely cramped conditions. This

Example of edge preparation for single vee butt weld on a larger diameter pipe.

is where the overhead welding technique becomes essential. As can be imagined, this is not only very difficult, but often dangerous, with the risk of blobs of molten metal falling on to the operator's clothing, or his face, or worse still, on his eyes or ears. Obviously, every precaution has to be taken by wearing the regulation leather apron, gauntlets, and some kind of protective covering for the face. The hand type of screen is usually found more convenient in these circumstances because of the limited space in which to maneuver, though many welders may still prefer the helmet type to which they are accustomed.

However, the novitiate welding operator is hardly likely to have to cope with this kind of specialized welding work, and I have only referred to it in order to give some indication of what may be anticipated should he at a later stage be attracted to this branch of welding by the undoubted high (and well-earned!) monetary rewards offered.

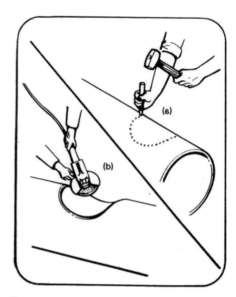

Fig. 49 *Preparing the horizontal component of a saddle joint.*

OTHER IMPORTANT JOINTS IN PIPE WELDING

The Tee or Saddle Joint. Before deciding on procedure, several factors must be considered, for example what will the particular pipe have to carry? Water? (Hot or cold). Oil? Or gas, perhaps? Or it could be any one of the many diversified uses to which tubes and pipes are today put. Also needed is the amount of pressure per square inch to which the pipe or tube may be subjected. Obviously, also where a pipe or tube is intended to carry a fluid of any kind, there must be no undue penetration or protuberance of the weld metal through to the inside of the pipe,

as this would cause an impediment to the flow. This calls for skill and care on the part of the operator, bearing in mind that while only single-vee preparation for the joint can be used, at the same time 100% penetration must be achieved. In the case of the saddle (or tee) joint, a great deal depends on whether or not a free flow of fluids is required to pass through the assembled joint. If the pipe is simply part of a tubular structure and is not intended to convey any fluids, etc. it is only necessary to cut the end of a vertical section to fit over the horizontal section, without any hole needing to be cut in the latter.

If, however, there has to be a free flow of the contents through the system of tubes and pipes, then a perfectly fitting aperture must first be cut into the horizontally-placed pipe, exactly conforming to the size and contour of the tee (or vertically placed) pipe. To achieve this, both pipes must be carefully marked off and for this purpose most up to date engineering firms have complete sets of metallic (tinfoil) templates of all shapes and sizes and with the use of these, marking off is made comparatively easy. It is a simple matter to wrap the correct flexible metallic template around the pipe and mark off the contour with a scriber. The same procedure is applicable for marking off the contours of the ends of pipes to be welded.

Marking the pipe joint faces without a template is a little tricky, the more difficult being the vertical pipe. It is possible to coat it with whitening or chalk (or, better, engineers' marking fluid) and to use a short length of the horizontal cross-pipe to guide a scriber to the required shape, or the two pipes may be held together and the shape plotted using a pair of dividers or a height gauge, if available. It may be possible to tape a sheet of card or tinplate round it, extending the tube so formed beyond the end of the pipe and snipping until it fits the cross-pipe, then sliding it back on the pipe and scribing round the fitted shape. The marking must be made more permanent by center-punching about every 42in. since when cutting with the oxy-flame a fine scribed line is likely to be burned away.

Once the vertical pipe is cut to a close fit on the cross-pipe it is only necessary to hold it in position and scribe round it mark the aperture required in the crosspipe. This line, too, must be center punched before applying the cutting flame. A neat and workmanlike fit of the two pipes will simplify the welding and help to achieve a 100% leak-proof joint.

In this connection it is worth passing on a useful tip from personal experience.

Even with the utmost care it is still possible that a leak may be discovered in, say, the pipework of the central heating system for a large block of offices or flats, and the leak may well be at an inaccessible place for welding, like at the back of a pipe which itself is fixed bang up against a wall. To weld and seal such a leak may seem impossible, but there is an easy if dramatic answer. Simply cut a hole in the front of the pipe opposite the leak just large enough to allow manipulation of the electrode (or gas blowpipe) and weld up the offending leak. Replace the cut-out piece of pipe and weld it back into place and the job is done.

Fig. 50 *Edge angles for a saddle joint. Weld opposite quarter circles.*

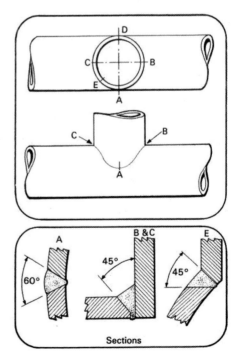

ARC WELDING STAINLESS STEEL PIPE

Stainless steel pipes present one rather difficult problem – it is not possible to cut stainless steel by the normal oxy-flame process and other means must therefore be used. Ordinary square ended cuts can be made with a power saw, or by hand with a hacksaw, or in the lathe. The edge preparation can then be carried out by grinding, or 'scurfing' as it is popularly termed, by a hand-held power grinder, or in the last resort by filing. Square cuts, then, are fairly simple. But what about in the case of welding a stainless steel branch pipe on to a main pipe? These can be marked out using the same methods as with mild steel, i.e. by using the appropriate template, with the aid of the scriber, or as described a few lines back. Now we are presented with the problem of cutting out the aperture in the main pipe, the shape of which must conform to that of the branch pipe at the joint. Stainless steel *can* be cut by the air-arc process – which is simply the ordinary metal-arc electrode accompanied by a high pressure jet of air – but this gives only a very roughly contoured cut, with very jagged edges, and not all engineering workshops have this sort of equipment. If this method is used, the rough edges and slag must be thoroughly cleaned off by grinding, etc. Where this system is not available, as a rough and ready method, an ordinary metal arc electrode can be used with a very high amperage, but only in the case of light fairly thin pipe, and again this gives a very rough cut, involving a great deal of cleaning and grinding.

However, there is another method. This is to drill a series of small holes along the scribed line, as closely together as possible, after which the jagged metal between the holes can be chipped away by hammer and chisel and the edges beveled off in the usual way, by grinding or filing. Then, from this stage onwards, the technique of arc welding of stainless steel pipes is basically the same as that of mild steel, except that it involves the use of special electrodes which must be of the correct type, appropriate to whatever stainless steel alloy is being used.

74

Chapter 10

Vertical and Overhead Arc Welding of Mild Steel

These welds are very difficult to carry out. Indeed, only very experienced and highly skilled welders can do these kinds of welding, but they play a very important part in welding work. To take vertical welding first. This occurs to a great extent in ship repair and shipyard work, and pipework. To conform with Department of Trade and Industry regulations, all vertical runs, or beads, with the exception of the root run, must be carried out by the Upward Vertical technique. This art can only be acquired by long practice. After the root run is completed (see below) all subsequent runs must be started from the bottom of the vee and continued upwards and each run or bead must overlap the previous one.

Single Vee Vertical Butt Weld. The first, or root run, can be done in the downward direction, by using cellulosic electrodes, in order to obtain adequate penetration. This should penetrate right through the material – and far enough through to project at the back of the vee, to enable a sealing run to be deposited there. In the case of material over, say, 55in. (3mm) thick one or more further runs (or beads) may be

necessary. The second run will have to be carried out using the weaving technique, in which the electrode (now of the normal type for this and all succeeding runs) is moved from side to side, as shown in Fig. 16. In the case of thicker material, where a third run is necessary, use a much wider weave, wide enough to ensure complete fusion with the parent metal at each side of the weld. In the case of extremely thick material, a series of multi-run beads must be deposited in the correct order. After completion of the vee and after thorough chipping and cleaning of the back of the weld, a sealing or capping run must be deposited.

With vertical welding, it is important to get a good start. Therefore, first fully establish the arc at the bottom of the joint, obtaining a good pool of molten metal there before moving upwards with a weaving motion, keeping the electrode pointing upwards at an angle of 70° – 80°.

Upward Vertical Welding of Corner Joints. Here again, the procedure is much the same as above, with the proviso that the weld must be built up to the full sectional thickness of the

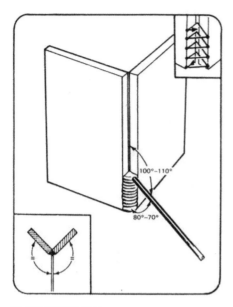

Fig. 51 *Vertical corner joint.*

has started to flow away from the molten pool of metal, it will continue to do so throughout the length of the welding joint. As an aid to this, it is a good idea to step up the current a little above normal at first, gradually reducing it as the weld proceeds.

Overhead – Single Vee Butt Weld on M/S of about 48in. (10mm) thickness. After beveling each edge 30° as usual, allowing a gap of $\frac{1}{16}$in. (2mm) tack in the usual way, spaced about 2 – 3ins. apart. Then commence as described above, holding the electrode in an almost vertical position (i.e. 90° to angle of welding joint) and deep in to the root of the vee, to ensure deep and full penetration. For further runs, the weaving technique must be employed, and, as for all welds, in the final run, full fusion must be maintained with the parent metal at each side of the weld. And at the risk of wearying the reader by constant repetition of this theme, the author makes no apology for once more referring to the vitally important task of chipping and thorough cleaning of the slag after each run, before proceeding with the next one.

Overhead Corner Weld. The root run for this can be done using cellulosic electrodes, but these are not essential. Actually cellulosic electrodes are specially designed for the downward vertical root run on large circular storage tanks, large diameter pipes, and similar fabrications; for all ordinary normal overhead and vertical welding, the normal type of electrode, such as the Vodex, is perfectly suitable.

The technique of carrying out an overhead corner weld is much the same as in the previous welds. The essential points are that after completing the root run, the next and all succeeding runs

material. Many operators fail to do this with the result that the weld is not up to full strength, (Fig. 39) and therefore leaves a weak point in the structure. The completed corner weld should appear as shown in Fig. 51.

Overhead (Metal-arc) Welding. In the overhead position, the operator has to work against the pull of gravity, but he will find that the electric arc can overcome this. The secret is to keep the arc very short, in fact using the electrode almost as a contact rod. B.O.C.'s Vodex electrodes are eminently suitable for either overhead or vertical positions.

In overhead welding, a good start is all-important. Thoroughly establish a good pool of molten metal right from the beginning, even allowing a little of this to over-run the edge to start with, especially the slag, as again, once this

76

must overlap the previous one, and in the final run, the material must be built up to the full strength of the parent metal, as with all corner welds, by making sure of full fusion with the parent metal on each side along the whole length of the joint.

Horizontal Welds on Vertical Surfaces.
This sounds like a contradiction in terms, but it is a technique which is sometimes necessary, for instance when a fabrication cannot be turned over for welding in the flat position but can only be stood up vertically. A good example is the bulkhead of a ship as it is being built, which has to be worked on in the vertical position. The edges of the mild steel plates to be joined, usually about 48in. (9mm) in thickness, are prepared in the usual way as for a single vee butt weld, with a gap of $^1/_{16}$in. (1.5 to 1.75mm) except that the lower face of the bevel is ground to only 15°, and the upper face to 45°. Both ends are tacked before commencing the root run. For this a very short arc is used, with the electrode held at about 65° from the horizontal, and without any weaving. The second run is deposited so that it overlaps the root run, and likewise with the third run. If this weld is being treated as a single vee the reverse side then has to be chipped and thoroughly cleaned and a sealing run deposited. Possibly, however, two welders may be employed, one on each side of the bulk-

Horizontal-vertical arc welding, typical of large storage tanks, ships' sides etc.

head, in which case the weld would be treated as a single vee butt weld at each side of the joint, with the edges prepared in a similar way, and of course no sealing run would be necessary.

Chapter 11

Building up and Reinforcement

Arc welding is valuable as a means of building up under-sized or damaged parts, e.g. worn parts of machinery, worn-down shafting, axles, etc., and especially valuable in the building up of worn teeth on machines such as excavators. The latter process requires special hard steel electrodes and is known as 'hard-facing'. For this B.O.C.'s 'Armex' electrodes are very well suited.

Let us take a simple example of reinforcement on 57in. mild steel plate,

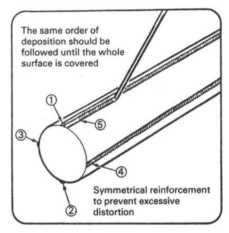

The same order of deposition should be followed until the whole surface is covered

Symmetrical reinforcement to prevent excessive distortion

Fig. 53 *Surfacing a shaft.*

Fig. 52 *Hard facing preparation.*

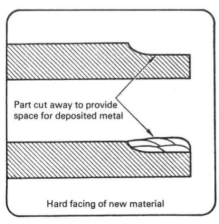

Part cut away to provide space for deposited metal

Hard facing of new material

in the down-hand (flat) position. After the first bead of weld deposit, the following bead must overlap the first, from its center, and likewise each succeeding run, until the whole piece of plate is uniformly covered. After this is done, we turn the plate round 90° and deposit the next layer at right angles to the first one, again each run overlapping the previous one. Further layers

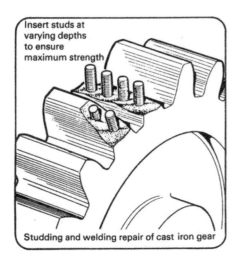

Insert studs at varying depths to ensure maximum strength

Studding and welding repair of cast iron gear

Fig. 54 *Repairing a cast iron gear.*

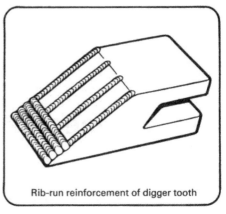

Rib-run reinforcement of digger tooth

Fig. 55 *Rib-run reinforcement.*

required. After completion, the surface may be ground, thus leaving a smooth finish.

Chapter 12

Resistance Welding

This form of welding is used extensively in sheet metal work, for which it is well adapted, particularly in spot welding, in a spot welding machine. This is almost a purely mechanical operation, in which very little skill is required. Spot welding is used extensively in car factories and in many other manufacturing shops requiring large numbers of sheet metal joints. The two pieces to be joined are inserted in position into the machine, the time switch is set, the current is switched on, the foot pedal pressed by the operator and the two pieces of sheet metal are fused together at one spot between the upper and lower electrodes. All the skill that is needed is to set the correct timing and the correct amperage; and the machine does the rest, as the operator presses the foot pedal lever.

Smaller versions of these machines are also extensively used in motor car manufacture. Great numbers of them are suspended from the roof by coil springs and when an operative requires to use one he simply reaches up, pulls down the appropriate welding 'gun', inserts it into the required spot, presses the lever, and the job is done; it can be repeated all along the seam if needed. The aforementioned Flash-welding is similarly another form of resistance welding.

British Rail and other railways also use resistance welding for joining their all-welded rails, using a special machine for the purpose. The two rails are placed in the machine end to end, and they are forced together under tremendous pressure as the current is switched on. The rail then becomes one whole, requiring only the slag and surplus metal to be removed to result in a perfectly finished job.

Chapter 13

T.I.G. and M.I.G. Welding

Electric Arc Welding Aluminum

Welding of aluminum by the oxyacetylene process has already been dealt with so let us take a look at the pros and cons of welding this metal by the electric-arc process. The great advantage this method possesses is, of course, speed, which is about six times greater than that of the gas-welding method.

Aluminum, and most of its alloys, can be electrically welded by using coated aluminum electrodes, but it is only possible by using D.C. current, obtained usually from an electric generator or rotary converter. Because of this the procedure is not in very general use today, and in fact it has been almost superseded in recent years by the much superior process known as T.I.G. (Tungsten Inert Gas) or M.I.G. (Metallic Inert Gas), or as sometimes referred to, argon-arc or CO2 welding.

T.I.G. and M.I.G. Systems

These systems of welding aluminum both have one advantage over all other methods of welding aluminum in that no flux whatever is required. The fluxes used in gas welding aluminum dissolve the film of oxide which is inescapably present on the surface of aluminum and its alloys. However much the surface is scrubbed, abraded with wire brushes or even filed, the oxide reforms almost immediately. Flux deals with this but carries the drawback that all traces must be removed after welding; the process is time-consuming, since the only certain method of complete removal involves immersing the welded work in hot water, followed by a quick dip in caustic soda, then a brief submersion in dilute nitric acid and finally thorough washing in cold water.

T.I.G. or M.I.G. welding eliminate the problem completely, by excluding atmospheric oxygen and nitrogen, etc. from the weld area, using an electric arc shielded or enveloped by an inert gas. In the first system, a tungsten electrode is used which does not burn away or melt into the weld like a normal electrode. It can be regarded as more or less permanent, although the tip tends to become pitted after prolonged use and has to be trued up by light grinding. It must also be kept clean and free from adhering spots of weld metal. The correct size tungsten electrode and appropriate size porcelain nozzle must be used for different material thickness-

es and the current and gas pressure are adjusted to suit. For fairly light work the cables and holder are air cooled, but heavier work requires them to be water cooled.

The difference with the M.I.G. method is that the electrode is a continuous length of welding wire fed mechanically from an overhead spool and geared to be fed at a rate synchronized with the welding speed. There is thus a continuous flow of welding feed wire into the weld area, which is also shielded by inert gas; argon or carbon dioxide are the commonest gases used. The gas has a scavenging effect which replaces the flux as a means of dissolving the film of oxide on the metal. An incidental advantage of the M.I.G. system is that it leaves the operator with a free hand, which enables time to be saved on light jobs when the operator can simply hold the work for tacking, dispensing with the need for clamps and holding devices.

Both methods involve more expensive equipment, but commercially this is compensated for by the elimination of the post-welding operation of flux removal and freedom from any danger of corrosion. With a skilled operator the welds resulting are beautifully formed and provide faultless fusion and, after grinding and polishing, invisible joins.

T.I.G., or argon-arc as it was initially called, has been in use since about 1946 and neither it nor M.I.G. is really a system for the amateur, though either should offer little difficulty to readers moving on to it. Use is not confined to aluminum and both systems are widely used for stainless steel and non-ferrous alloys. In the U.S.A. they are extensively used (with a small amount of oxygen added to the gas) for mild steel, but at present this is not considered a sufficiently economic proposition in Britain.

Edge Preparation. Edge preparation for single and double-vee butt welds of aluminum by any of the above processes is to all intents and purposes identical with that for mild steel. That is, chamfering the edges to an angle of 30° on each edge for the single-vee joint, but leaving a slightly blunt edge at the bottom of the vee – not a knife edge, as this would burn away too quickly. For a

double-vee joint, edge preparation is the same, except that chamfering is carried out both top and bottom, but leaving a slightly blunt nose at the center. In both cases always leave a slight gap of about $1/_{16}$in. (1.6mm) in order to help achieve 100% penetration right through the material.

For fillet welds and tee joints no edge preparation is necessary, and the techniques for assembly and tacking are much the same as those for mild steel, and the same precautions should be taken to avoid distortion.

Aluminum has a very high thermal conductivity and a little preliminary preheating, though not absolutely essential,

Manual application of M.I.G. welding on an extensively jigged and clamped workpiece. Return can be seen clamped to base of steel bench.

always helps, especially on a very big job, involving a large mass of the metal. However, in dealing with light sheet metal aluminum, edge preparation is hardly necessary for thicknesses of less than 10 gauge (55in. or 3mm), though slight chamfering can be carried out if thought desirable. It is really not essential provided the $1/_{16}$in. gap is maintained between the two sections to be welded, as this performs the same function as a vee, provided the operator makes sure of 100% penetration when carrying out the weld.

Corner Welds on Aluminum, (as for example, on Box-shaped fabrications). As in mild steel fabrications of this kind, it is best first to place a tack about every inch, along the whole length of the corner joint, before proceeding with the weld proper. The help of a colleague may be needed to hold the two sections together while the operator is tacking. Place the two sheets of aluminum together with one edge very slightly overlapping the edge of the other, so as to leave a slight vee to accommodate the final weld. This is in fact very important, because if the edges are placed together with one completely overlapping the other, there is obviously then no room left for the weld deposit, and if welded in this position and the bead of weld metal later ground off then clearly there is nothing left to hold the two sections together. This principle applies to all corner welds, of whatever metal.

In T.I.G. welding of aluminum the technique is very similar to that of ordinary gas-welding, with a pool of metal first established and the filler wire fed into the pool of molten metal in just the same way.

Welding Stainless Steel by T.I.G. Process. The T.I.G. process is ideal for welding stainless steel, but there are many types of stainless steel, and the operator must know exactly which type he is working on, and also the correct filler rod to use for each type. To obtain this information, he must obtain guidance from the manufacturer. Argon gas is usually the most suitable, but if CO_2 is used, first seek expert confirmation that it is suitable for the particular type of stainless steel being worked on, because they and the parent metal vary so much in composition, with varying percentages of chromium, nickel, and so on. In the case of very thin sheet stainless steel, the edges may be slightly flanged with the depth of the flanges about $1/_{16}$in. (1.6mm) – tightly butted up together by clamping. With this assembly no filler rod is needed, the flanges simply being fused together, but right from the start the flanges must be fully melted until a full pool of molten metal is formed, before progressing to the end of the seam.

Technique of Vertical Corner Weld on Thin Stainless Steel Sheet. Use flanged edges in a similar way as before, clamping up in the usual way. This weld may be carried out by the downward vertical method, without using filler rod. Move the welding torch downward fairly fast, while at the same time completing fusion of the two edges. A little of the filler rod may be found necessary at the bottom end of the run, to build up the edge to normal thickness.

Overhead Lap Joint on Thin Stainless Steel Sheet. This again may be flanged, clamping together as before, then four tacks placed equally spaced along

Manual arc welding using the T.I.G. system in car production. The need for protective clothing is obvious.

the seam. Some filler rod may be necessary for this weld. To take the weight from the operator's arm, the cables should be slung over the shoulder and wrapped round the forearm, which helps maintain steadiness. The weld should be carried out by the leftward method, and with a slightly weaving motion, using filler rod as required, especially at end of run to ensure full build-up of the edge to the full strength of the parent metal. Any burrs should have been cleaned off the edges before commencing. On thicker material, two or more runs may be needed. If so, thoroughly clean the first bead of weld before proceeding with second or further runs. Clean with a stainless steel brush. Never use a mild steel wire brush, which is liable to leave small specks of mild steel embedded in the material and could cause rusting later. If through lack of a folding machine flanging is not pos-sible, thin sheet stainless steel can be T.I.G. welded by using a chill block or bar clamped to the metal sheets with the edges to be joined tight together. The chill bar helps to conduct the heat away. This method is suitable for thicknesses up to about 55in. (3mm) but over this thickness no chill bar will be necessary. Thicker material can be T.I.G. welded by clamping in the normal way, with the edges tightly together, and tacked in the usual way, i.e. place the first tack in the center, then use the back-stepping method, that is place the second tack to the right of the first one, and the third tack to the left of the first, and so on, along the length of the seam. This method avoids distortion, which would occur if all tacks were placed consecutively.

T.I.G. Welding of Heavier Sections of Stainless Steel. For thicknesses of over 55in. (3mm) the edges should be chamfered and treated as a single vee butt

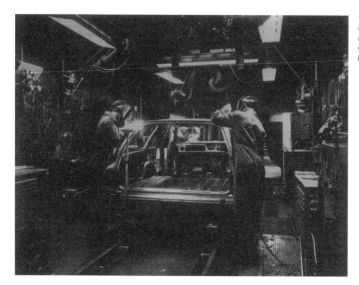

weld. Stainless steel plate of 48in. (9mm) should be treated as double vee butt welds. From this point, the procedure is the same as that for mild steel except that the correct filler rod must be used for whatever type of stainless steel is being welded.

Welding Copper by the Tungsten Inert Gas Process (T.I.G.). There are certain deoxidized grades of copper which are best suited for welding by T.I.G., notably grades C106 and C107. Other grades can be T.I.G. welded but are not so highly recommended for this, being liable to suffer from embrittlement as a result of the heat involved in the welding process. Argon, helium, or nitrogen may be used as shielding gases, but of these, helium results in the neatest welds. It is very important that the correct filler rod is used, according to the particular grade of copper being welded. With argon gas, the most suitable filler rod is of silicon

manganese deoxidized copper, i.e. as quoted above – grade 106 and 107.

As is well known, copper's most prominent characteristic is its very high thermal conductivity, or in simpler terms, it is a very good conductor of heat, and for this reason some preliminary pre-heating may be necessary, especially when argon gas (the gas in most general use) is being used.

Procedure for a Butt Weld on 1/16in. (1.6mm) Sheet Copper. First, remove all burrs from the cut edges and thoroughly clean off all grease and dirt. Also paste the joint area with a suitable flux to reduce oxidation. The flux may be the same as that used for the oxyacetylene welding of copper. This must be thoroughly cleaned off after welding.

Joint Assembly. Taking two pieces of sheet copper of the above thickness, clamp them edge to edge tightly together, or in a jig fixture. Tacks may be used, but they must be very strongly done with filler rod, as they are liable to crack during welding. If this happens, the tack must be re-welded with full fusion. When starting the weld proper commence with a good pool of molten metal and maintain this along the whole length of the joint, but do not weave the tungsten arc. At the end of the run, build up the edge of the material with plenty of filler rod.

METAL TRANSFER IN GAS-SHIELDED WELDING

For good penetration and high deposition rates, currents of 240-500A may be used with the spray transfer system **(Fig. 56)**, *where small droplets of molten metal from the electrode free-fly. This system can be used with aluminum in any position but only in the downhand position with other metals. For other positions*

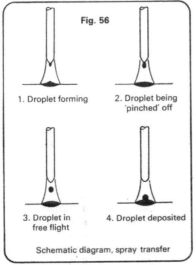

Fig. 56

1. Droplet forming   2. Droplet being 'pinched' off

3. Droplet in free flight   4. Droplet deposited

Schematic diagram, spray transfer

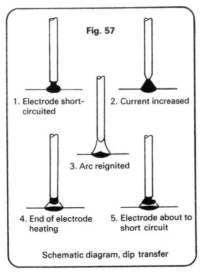

Fig. 57

1. Electrode short-circuited   2. Current increased

3. Arc reignited

4. End of electrode heating   5. Electrode about to short circuit

Schematic diagram, dip transfer

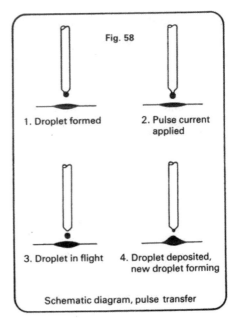

Fig. 58

1. Droplet formed   2. Pulse current applied

3. Droplet in flight   4. Droplet deposited, new droplet forming

Schematic diagram, pulse transfer

or thin material the current is reduced to 80-200A and metal transfer is by dip or pulse. In the former, used with a low current short arc, the end of the electrode touches the weld pool, current rises, metal melts off, electrode touches etc. 50-200 times per second, but this system is only suitable for metals with fairly high electrical resistance, such as steel **(Fig. 57)**. *In other cases a special machine supplies up to 100 high current pulses to the electrode* **(Fig. 58)** *produced metal transfer similar to spray transfer but at considerably lower average current.*

Weld Symbols

Engineering Working Drawings

Working drawings present views of a workpiece in three ways, plan, elevations and sections. The plan view is as seen from above, looking down vertically. Elevations are side and/or end views, and sections are drawings on planes cut through the object, vertically or horizontally. All large objects are drawn to a scale, which is shown on the drawing as a scale line with full-size dimensions marked as they appear reduced, or a statement, e.g. ¼in. – 1ft., or sometimes as a fraction (1/10 full size) or a ratio (1:25). However, drawings for constructional engineering include all dimensions and materials; usually non-critical dimensions are expressed in feet and/or inches and fractions (or, of course, meters) and important dimensions in decimal parts of an inch (or millimeters).

Symbols are used in welding drawings, as illustrated, and indicate the type of weld required (single or double butt, fillet, T joint, corner, lap, single or double U joints, etc.) plus, usually, dimensions of gaps, grinding angles, etc. It will be noted that the illustrations include one or two forms of joint not discussed in this book, but the uses are self-evident and all the symbols rapidly become familiar to a welder needing to work from such drawings.

Weld symbols for types of joints

BS 499 – Pt 2 – 1980

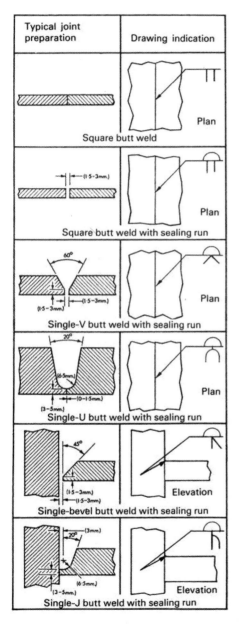

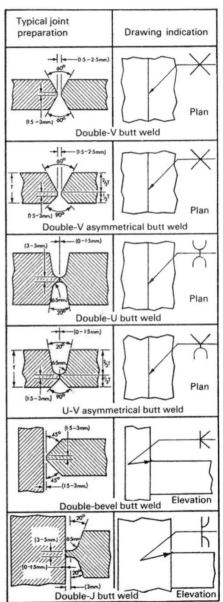

Typical joint preparation	Drawing indication

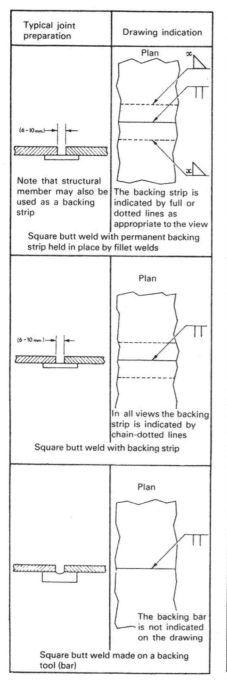

Left column, top row:

(6 – 10 mm)

Plan

Note that structural member may also be used as a backing strip

The backing strip is indicated by full or dotted lines as appropriate to the view

Square butt weld with permanent backing strip held in place by fillet welds

Left column, middle row:

(6 – 10 mm)

Plan

In all views the backing strip is indicated by chain-dotted lines

Square butt weld with backing strip

Left column, bottom row:

Plan

The backing bar is not indicated on the drawing

Square butt weld made on a backing tool (bar)

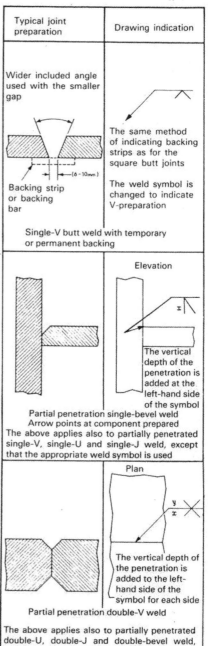

Typical joint preparation	Drawing indication

Right column, top row:

Wider included angle used with the smaller gap

(6 – 10 mm)

Backing strip or backing bar

The same method of indicating backing strips as for the square butt joints

The weld symbol is changed to indicate V-preparation

Single-V butt weld with temporary or permanent backing

Right column, middle row:

Elevation

The vertical depth of the penetration is added at the left-hand side of the symbol

Partial penetration single-bevel weld
Arrow points at component prepared
The above applies also to partially penetrated single-V, single-U and single-J weld, except that the appropriate weld symbol is used

Right column, bottom row:

Plan

The vertical depth of the penetration is added to the left-hand side of the symbol for each side

Partial penetration double-V weld

The above applies also to partially penetrated double-U, double-J and double-bevel weld, except that the appropriate weld symbol is used

Type of joint	Drawing indication

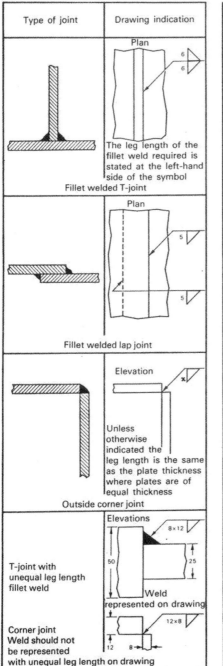

Fillet welded T-joint

The leg length of the fillet weld required is stated at the left-hand side of the symbol

Fillet welded lap joint

Outside corner joint

Unless otherwise indicated the leg length is the same as the plate thickness where plates are of equal thickness

T-joint with unequal leg length fillet weld

Corner joint
Weld should not be represented with unequal leg length on drawing

Weld represented on drawing

Type of joint	Drawing indication

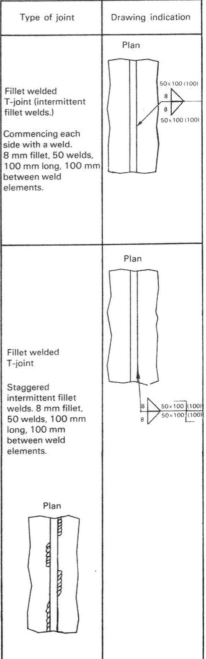

Fillet welded T-joint (intermittent fillet welds.)

Commencing each side with a weld. 8 mm fillet, 50 welds, 100 mm long, 100 mm between weld elements.

Fillet welded T-joint

Staggered intermittent fillet welds. 8 mm fillet, 50 welds, 100 mm long, 100 mm between weld elements.

Type of joint	Drawing indication
	Plan

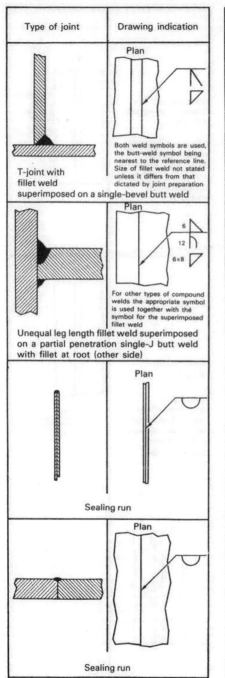

Supplementary instruction	Drawing indication
Weld all round a joint, eg. a flange to a pipe; a stanchion to a base-plate. A peripheral weld	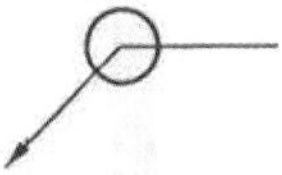 Placed at 'elbow' of arrow shaft with the reference line
To be welded on site	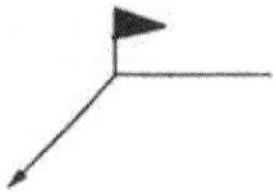
Flush finish to butt weld	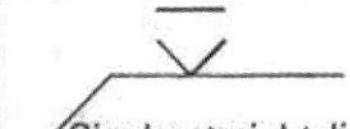 Single straight line added to symbol. This may be used with any type of butt weld with appropriate symbol, and may be used to request flush finish on one or both sides of the weld
Convex finish (to butt weld)	
Concave finish (to fillet weld)	
Weld to be radiographed	 NDT Symbol is to attract attention, added at end of reference line bearing appropriate weld symbols

Left table image text:

Both weld symbols are used, the butt-weld symbol being nearest to the reference line. Size of fillet weld not stated unless it differs from that dictated by joint preparation

T-joint with fillet weld superimposed on a single-bevel butt weld

For other types of compound welds the appropriate symbol is used together with the symbol for the superimposed fillet weld

Unequal leg length fillet weld superimposed on a partial penetration single-J butt weld with fillet at root (other side)

Plan

Sealing run

Plan

Sealing run

Aids to Erection and Assembly

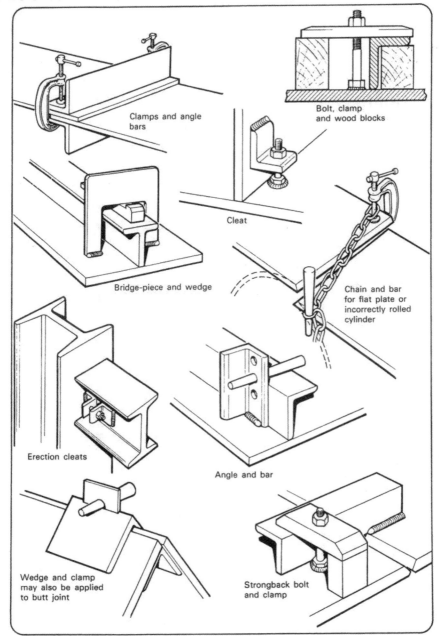

Clamps and angle bars

Bolt, clamp and wood blocks

Cleat

Bridge-piece and wedge

Chain and bar for flat plate or incorrectly rolled cylinder

Erection cleats

Angle and bar

Wedge and clamp may also be applied to butt joint

Strongback bolt and clamp

Welding Defects and most likely causes

1 Unequal leg length fillet. *Incorrect angle of filler rod and blowpipe.*

2 Fillet weld with insufficient throat thickness. *Speed of travel too fast, leading to insufficient deposited weld metal.*

2a Fillet weld with excessive throat thickness. *Speed of travel too slow causing heavy deposit. Filler rod too large.*

3 Excessive concavity in butt weld profile. *Excess heat build-up with too fast a speed of travel, or filler rod too small.*

4 Excessive convexity in butt weld profile. *Insufficient heat – too slow a speed of travel – nozzle size too small – filler rod too large.*

5 Undesirable weld profile (lap fillet) – excess melting of plate edge, giving insufficient throat thickness. *Incorrect tilt angle of blowpipe fusing top edge of plate, which flows down to produce unequal leg length fillet with undesirable profile.*

6 Notch effect with overlap at side of fillet weld. *Incorrect manipulation together with incorrect angle of blowpipe and filler rod.*

7 Notch effect with overlap at side of butt weld. *Incorrect manipulation together with incorrect angle of blowpipe and filler rod.*

8 Excessive penetration. Excess fusion of root edges. *Angle of slope of nozzle too large. Insufficient forward heat. Flame size and/or velocity too high. Filler rod too large or too small. Speed of travel too slow.*

9 Burn through. *Excessive penetration. has produced local collapse of weld pool resulting in a hole in the root run.*

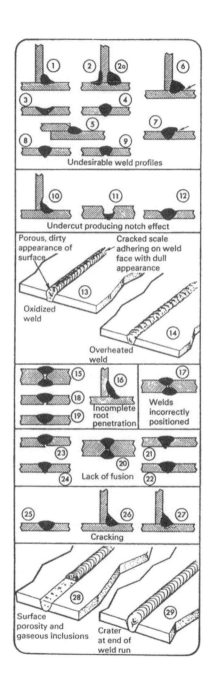

Undesirable weld profiles

Undercut producing notch effect

Porous, dirty appearance of surface

Cracked scale adhering on weld face with dull appearance

Oxidized weld

Overheated weld

Incomplete root penetration

Welds incorrectly positioned

Lack of fusion

Cracking

Surface porosity and gaseous inclusions

Crater at end of weld run

10 Under-cut along vertical member of fillet welded tee joint. *Incorrect angle of tilt used in blowpipe manipulation.*

11 Root run too large with undercut in butt joint. *Use of too large a nozzle and/or excessive lateral blowpipe manipulation with too slow a speed of travel.*

12 Under-cut both sides of weld face in butt joint. *Wrong blowpipe manipulation; incorrect distance from plate surface; excessive lateral movement. Use of too large a nozzle.*

13 Oxidized weld face. *Use of oxidizing flame setting. Insufficient cleaning of plate surfaces. Incorrect manipulation of blowpipe permitting cone to contact the molten pool. Atmospheric contamination.*

14 Overheated weld. *Use of too large a nozzle. Speed of travel too slow. Excess blowpipe manipulation extending the weld pool.*

15 Incomplete root penetration in butt joints (single vee or double vee). *Incorrect set-up and joint preparation. Use of unsuitable procedure and/or welding technique.*

16 Incomplete root penetration in close square tee joint. *Incorrect set-up and joint preparation. Use of unsuitable procedure and/or welding technique.*

17 Welds incorrectly positioned. *Welds have been deposited out of alignment with the center line of the joint.*

18 Notch, instead of root underbead. Lack of root penetration. *Angle of nozzle too small. Speed of travel too fast. Insufficient heat applied.*

19 Lack of root penetration. *Incorrect joint penetration and set up. Gap too small. Vee preparation too narrow. Root edges touching.*

20 Lack of fusion on root and side faces of double vee butt joint. *Incorrect set-up and joint preparation. Use of unsuitable welding technique.*

21 Lack of fusion along lower edge of one side face in single vee butt joint. *Incorrect technique. Angle of blowpipe tilt. Concentrating heat mainly on one fusion face.*

22 Lack of inter-run fusion. *Angles of nozzle and blowpipe manipulation incorrect.*

23 Lack of fusion of one or both root edges with lack of root penetration. *Incorrect alignment of joint edges.*

24 Lack of fusion between deposited metal and root edges (lack of root fusion). *Slope of nozzle too small. Too much heat dissipated forward and filler rod melts too soon.*

25 & Weld face cracks in butt and fillet welds.
26 *Use of incorrect welding procedure. Unbalanced expansion and contraction stresses. The presence of impurities. Undesirable chilling effects. Use of incorrect filler rod.*

27 Toe and underbead cracks. *Use of incorrect procedure leading to internal stresses. Chilling effects producing hardening of parent metal in thermally disturbed zone.*

28 Surface porosity and gaseous inclusions. *Use of incorrect filler rod and technique. Failure to clean surfaces before welding. Absorption of gases. Incorrectly stored fluxes, unclean filter rod. Atmospheric contamination.*

29 Crater at end of weld run. Small cracks may be present. *Neglect to change the angle of blowpipe, speed of travel or increase the rate of weld metal deposition as welding is completed at the end of the seam.*

(By courtesy of the Engineering Industry Training Board, Watford)